Glencoe McGraw-Hill

Grade 8

Math
Triumphs

Book 3: Data ..., Number ...rations, ...ebra

Authors
Basich Whitney • Brown • Dawson • Gonsalves • Silbey • Vielhaber

McGraw Hill Glencoe

Photo Credits

All coins photographed by United States Mint.
All bills photographed by Michael Houghton/StudiOhio.
Cover Jupiterimages; **iv** (1 7 8)File Photo, (2 3)The McGraw-Hill Companies, (4 5 6)Doug Martin; **vi, vii** CORBIS; **viii** CORBIS; **viii** PunchStock; **272–273** Jupiterimages; **284** Getty Images; **286** Ed Kashi/CORBIS; **293** Kemberly Groue/U.S. Air Force; **334** AGE Fotostock; **340–341** Getty Images; **344** Ken Cavanagh/The McGraw-Hill Companies; **346** PunchStock; **347** CORBIS; **382–383** Masterfile; **388** (t)G.K. & Vikki Hart/Getty Images, (b)Getty Images; **390** Kent Knudson/Getty Images.

The *McGraw·Hill* Companies

Macmillan/McGraw-Hill
Glencoe

Send all inquiries to:
Glencoe/McGraw-Hill
8787 Orion Place
Columbus, OH 43240-4027

ISBN: 978-0-07-888215-9
MHID: 0-07-888215-X

Math Triumphs
Grade 8, Book 3

Printed in the United States of America.

3 4 5 6 7 8 9 10 066 17 16 15 14 13 12 11 10 09

Math Triumphs

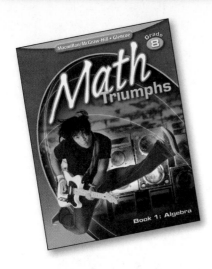

Book 1

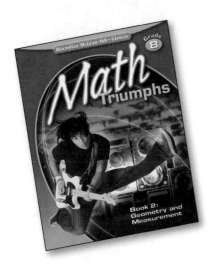

Book 2

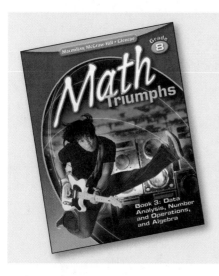

Book 3

Authors and Consultants

AUTHORS

Frances Basich Whitney
Project Director, Mathematics K–12
Santa Cruz County Office of Education
Capitola, California

Kathleen M. Brown
Math Curriculum Staff Developer
Washington Middle School
Long Beach, California

Dixie Dawson
Math Curriculum Leader
Long Beach Unified
Long Beach, California

Philip Gonsalves
Mathematics Coordinator
Alameda County Office of Education
Hayward, California

Robyn Silbey
Math Specialist
Montgomery County Public Schools
Gaithersburg, Maryland

Kathy Vielhaber
Mathematics Consultant
St. Louis, Missouri

CONTRIBUTING AUTHORS

Viken Hovsepian
Professor of Mathematics
Rio Hondo College
Whittier, California

FOLDABLES Study Organizer **Dinah Zike**
Educational Consultant,
Dinah-Might Activities, Inc.
San Antonio, Texas

CONSULTANTS

Assessment

Donna M. Kopenski, Ed.D.
Math Coordinator K–5
City Heights Educational Collaborative
San Diego, California

Instructional Planning and Support

Beatrice Luchin
Mathematics Consultant
League City, Texas

ELL Support and Vocabulary

ReLeah Cossett Lent
Author/Educational Consultant
Alford, Florida

Reviewers

Each person below reviewed at least two chapters of the Student Edition, providing feedback and suggestions for improving the effectiveness of the mathematics instruction.

Patricia Allanson
Mathematics Teacher
Deltona Middle School
Deltona, Florida

Debra Allred
Sixth Grade Math Teacher
Wiley Middle School
Leander, Texas

April Chauvette
Secondary Mathematics Facilitator
Leander Independent School District
Leander, Texas

Amy L. Chazarreta
Math Teacher
Wayside Middle School
Fort Worth, Texas

Jeff Denney
Seventh Grade Math Teacher, Mathematics
    Department Chair
Oak Mountain Middle School
Birmingham, Alabama

Franco A. DiPasqua
Director of K-12 Mathematics
West Seneca Central
West Seneca, New York

David E. Ewing
Teacher
Bellview Middle School
Pensacola, Florida

Mark J. Forzley
Eighth Grade Math Teacher
Westmont Junior High School
Westmont, Illinois

Virginia Granstrand Harrell
Education Consultant
Tampa, Florida

Russ Lush
Sixth Grade Math Teacher, Mathematics
    Department Chair
New Augusta - North
Indianapolis, Indiana

Joyce B. McClain
Middle School Math Consultant
Hillsborough County Schools
Tampa, Florida

Suzanne D. Obuchowski
Math Teacher
Proctor School
Topsfield, Massachusetts

Karen L. Reed
Sixth Grade Pre-AP Math
Keller ISD
Keller, Texas

Deborah Todd
Sixth Grade Math Teacher
Francis Bradley Middle School
Huntersville, North Carolina

Susan S. Wesson
Teacher (retired)
Pilot Butte Middle School
Bend, Oregon

Contents

Chapter 7 · One-Variable Data

Surfing in Hawaii

Chapter 8 Percents and Circle Graphs

Bryce Canyon National Park, Colorado

Contents

Chapter 9 — Two-Variable Data

Jonathan Dickinson State Park, Florida

SCAVENGER HUNT

BOOK 1

Let's Get Started

Use the Scavenger Hunt below to learn where things are located in each chapter.

1. What is the title of Lesson 9-3?

2. How many examples are provided in Lesson 7-3?

3. On what page can you find the vocabulary term *percent* in Lesson 8-1?

4. What are the vocabulary words for Lesson 7-2?

5. How many exercises are presented in the Chapter 8 Study Guide for Lesson 8-2?

6. What strategy is used in the Step-by-Step Problem-Solving Practice on page 285?

7. List the operations that are shown in Example 1 on page 289.

8. Describe the art on page 296 that accompanies Exercise 6.

9. On what pages will you find the Test Practice for Chapter 7?

10. In Chapter 8, find the logo and Internet address that tells you where you can take the Online Readiness Quiz.

Chapter 7

One-Variable Data

How are survey results displayed?

Roberta surveyed her friends to find out which animal was the most popular pet choice. Ten friends chose cats and twelve friends chose dogs. Three friends liked both cats and dogs equally. How could Roberta display this information?

STEP 1 Quiz

Math Online Are you ready for Chapter 7? Take the
Online Readiness Quiz at *glencoe.com* to find out.

STEP 2 Preview

Get ready for Chapter 7. Review these skills and compare
them with what you will learn in this chapter.

What You Know	What You Will Learn
You know how to add and how to divide.	*Lesson 7-3*

What You Know

You know how to add and how
to divide.

Examples:

$4 + 4 + 8 + 12 = 28$

$28 \div 4 = 7$

TRY IT!

1 $7 + 9 + 2 + 18$ **2** $36 \div 9$

_____   _____

3 $5 + 9 + 2 + 4$ **4** $20 \div 5$

_____   _____

What You Will Learn

Lesson 7-3

The **mean** is the sum of the numbers
in a set of data divided by the
number of pieces of data.

Enrique recorded the ages of
his cousins.

2, 5, 7, 3, 9, 4

To find the mean age of Enrique's
cousins, add the ages.

$2 + 5 + 7 + 3 + 9 + 4 = 30$

Then divide by the number of pieces
of data (the number of Enrique's
cousins).

$30 \div 6 = 5$

The mean age is 5.

What You Know

You know how to compare values in a
set of numbers.

47, 96, 64, 18, 73

The smallest value is 18.
The largest value is 96.

TRY IT!

5 90, 97, 93, 95, 99, 90

The smallest value is _____.
The largest value is _____.

What You Will Learn

Lessons 7-5 and 7-7

Both bar graphs and line graphs use
a **scale** to compare data.

The **interval** of the scale is
determined by the **range** of the data.

37, 51, 43, 4, 72, 18

The numbers are between 4 and 72.

Since $72 - 4 = 68$, the range of the
data is 68.

A reasonable scale would use an
interval of 10, and the scale should
extend from 0 to 80.

Sort and Classify

KEY Concept

A **Venn diagram** is one way to show how objects and numbers are sorted.

Follow these steps to sort and classify objects.
1. Sort objects by attribute.
2. Use a Venn diagram to show how the objects are sorted.

These objects are sorted by color and by shape.

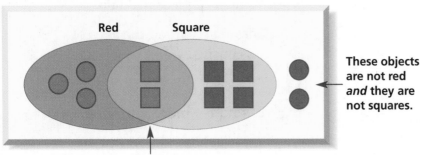

These objects are red *and* they are squares.

These objects are not red *and* they are not squares.

VOCABULARY

sort
to put together items that have something in common

Venn diagram
a diagram that uses circles to display elements of different sets; overlapping circles show common elements

Venn diagrams can show common features among three categories. This diagram shows the factors of 18, 24, and 32.

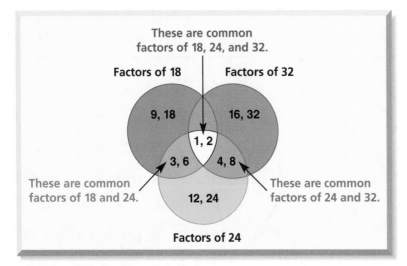

These are common factors of 18, 24, and 32.

Factors of 18

Factors of 32

9, 18

16, 32

1, 2

3, 6

4, 8

These are common factors of 18 and 24.

These are common factors of 24 and 32.

12, 24

Factors of 24

1 and 2 are common factors of 18, 24, and 32.

To sort and classify, think about how the objects are alike. Put the items that are alike in a group.

Example 1

Create a Venn diagram to sort the numbers. Classify them as even numbers or as squares of whole numbers.

1, 2, 3, 4, 5, 6, 7, 8, 9

1. Sort and classify the numbers.

Even:	**Squares of Whole Numbers:**	**Neither:**
2, 4, 6, 8	4, 9	1, 3, 5, 7

2. Use a Venn diagram to show how the numbers are sorted. Identify the numbers in each group, in both groups, and in neither group.

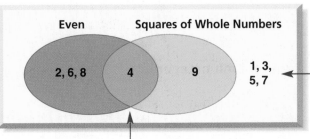

These numbers are not even *and* are not squares of whole numbers.

These numbers are even *and* squares of whole numbers.

YOUR TURN!

Create a Venn diagram to sort the numbers. Classify them as odd numbers or multiples of 3.

20, 21, 22, 23, 24, 25, 26, 27, 28, 29, 30

1. Sort and classify the numbers.
   Odd:

   _____, _____, _____, _____, _____

   Multiples of 3:

   _____, _____, _____, _____

   Neither:

   _____, _____, _____, _____

2. Use a Venn diagram to show how the numbers are sorted. Identify the numbers in each group, in both groups, and in neither group.

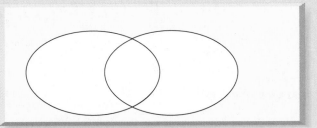

GO ON

Who is Correct?

What is the title for the category of numbers that contains 1 and 9?

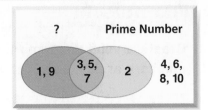

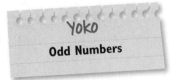
Olivia
Composite Numbers

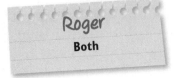
Yoko
Odd Numbers

Roger
Both

Circle correct answer(s). Cross out incorrect answer(s).

 Guided Practice

Sort the numbers 1, 2, 3, 4, 5, 6, 7, 8, 9, and 10 into each category.

I whole number factors of 12:

_____

2 even numbers:

_____

Step by Step Practice

3 Create a Venn diagram to sort the numbers. Classify them as multiples of 10 or two-digit numbers.

$$50, 67, 80, 93, 106, 110, 125, 200$$

Step 1 Sort and classify the numbers.

Multiples of 10:

_____

Two-Digit Numbers:

_____

Neither:

_____

Step 2 Create a Venn diagram.

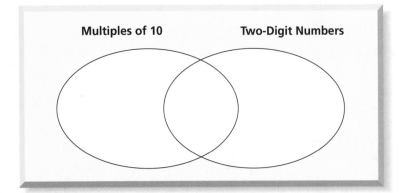

4 Create a Venn diagram to sort the numbers **15, 25, 30, 45, 65, 80, 90** and **100**. Classify them as multiples of 15 or multiples of 10.

Multiples of 15:   Multiples of 10:   Neither:

_____   _____   _____

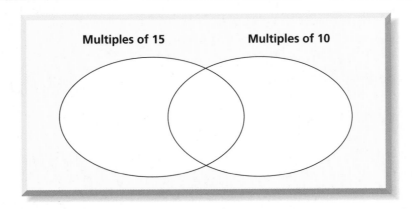

Multiples of 15   Multiples of 10

Step by Step Problem-Solving Practice

Solve.

5 **FOOD**   Amado classified apples, grapes, pears, oranges, bananas, strawberries, pineapples, and raspberries into two categories. Which fruits did Amado put in both categories?

Problem-Solving Strategies
- ☑ Draw a diagram.
- ☐ Look for a pattern.
- ☐ Guess and check.
- ☐ Act it out.
- ☐ Solve a simpler problem.

Understand   Read the problem. Write what you know. You need to find fruits that can be peeled or do not have to be peeled.

Plan   Pick a strategy. One strategy is to draw a diagram. Create a Venn diagram to classify types of fruit.

Solve

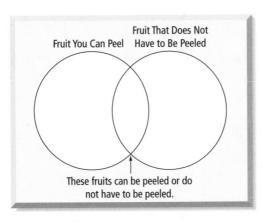

Fruit You Can Peel   Fruit That Does Not Have to Be Peeled

These fruits can be peeled or do not have to be peeled.

_____ are in both categories.

Check   Is your answer reasonable?

GO ON

6 **NUMBER SENSE** Curt noticed that some numbers were made of straight lines and some had curved lines.

0 1 2 3 4 5 6 7 8 9

Which two numbers did Curt put in both categories?
Check off each step.

_____ **Understand: I underlined key words.**

_____ **Plan: To solve the problem, I will** _____.

_____ **Solve: The answers are** _____ **and** _____.

_____ **Check: I checked my answer by** _____

_____

7 **Reflect** List three ways to classify the numbers 5, 10, 15, 20, 25, and 30.

_____

_____

▶ Skills, Concepts, and Problem Solving

Sort each set of numbers into each category.

3, 7, 12, 14, 21, 32, 45, 50

8 multiples of 3: _____   **9** even numbers: _____

10 both: _____   **11** neither: _____

2, 3, 4, 17, 25, 50, 75, 90

12 factors of 100: _____   **13** odd numbers: _____

14 both: _____   **15** neither: _____

16 Create a Venn diagram to sort the numbers. Classify them as multiples of 8 or multiples of 10.

16, 20, 32, 35, 40, 50, 75, 80, 92

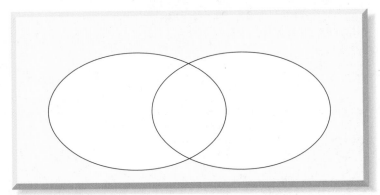

17 Create a Venn diagram to sort the numbers. Classify them as negative integers or even numbers.

−4, −3, −2, −1, 1, 2, 3, 4

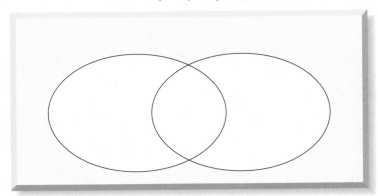

Solve.

18 **FOOD** During lunch, Betsy wrote down how many students chose green beans, potatoes, or both. Betsy made the Venn diagram below.

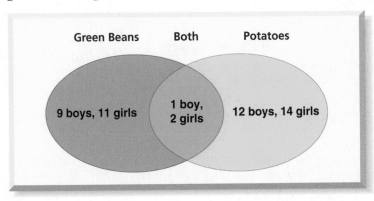

Green Beans | Both | Potatoes

9 boys, 11 girls | 1 boy, 2 girls | 12 boys, 14 girls

GO ON

Out of 49 students, how many chose both vegetables? _____

Vocabulary Check **Write the vocabulary word that completes each sentence.**

19 A(n) _____ is a diagram that uses overlapping and separate circles or ellipses to organize and show data.

20 **Writing in Math** Classify the numbers 3, 6, 7, 9, 12, 14, 21, and 28. Explain how to sort them.

_____

_____

21 Create a Venn diagram to sort the numbers. Classify them as even numbers or multiples of 5.

100, 101, 102, 103, 104, 105, 106, 107, 108, 109, 110

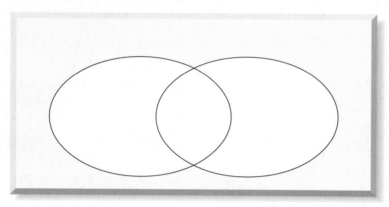

Solve.

22 **NUMBER SENSE** Donald listed all the multiples of 4 up to 50. Lia listed all the multiples of 6 up to 50.

Donald
4, 8, 12, 16, 20, 24, 28, 32, 36, 40, 44, 48

Lia
6, 12, 18, 24, 30, 36, 42, 48

The students made a Venn diagram of the lists. Which multiples were in both categories? Explain your answer.

_____

_____

_____

STOP

Mode, Median, and Range

KEY Concept

Data sets can be described by finding the mode, median, and range.

Five homerooms at Poncho Prairie Middle School collected quarters for hurricane relief. The coin jars are shown below.

| 117 | 125 | 209 | 174 | 117 |

The **mode** is the number that appears most often in a set of data. In this set of data the number 117 appears twice, so it is the mode.

The **median** is the number that falls exactly in the middle of a set of data. In order to find the median, the terms must be arranged in order from least to greatest, or greatest to least.

117, 117, (125) 174, 209

In a list of five terms, the third number will be the median. There are two numbers listed before the median and two numbers listed after the median. The median number of quarters collected is 125.

The **range** is the difference between the greatest and least number in a set of data. In this set of data, the greatest number of quarters collected is 209 and the least number of coins collected is 117. The range is 92.

209 − 117 = 92

VOCABULARY

data
    information, often numerical, which is gathered for statistical purposes

median
    the middle numbers in a set of data when the data are arranged in numerical order; if the data has an even number, the median is the mean of the two middle numbers

mode
    the number(s) or item(s) that appear most often in a set of data

range
    the difference between the greatest number and the least number in a set of data

Always put the data set in order from least to greatest before trying to describe the data set.

GO ON

Example 1

Find the mode for the given set of data.

Henry asked nine classmates how many DVDs they own.

$$4, 2, 0, 8, 6, 4, 5, 4, 0$$

1. Arrange the numbers in order from least to greatest.

$$0, 0, 2, 4, 4, 4, 5, 6, 8$$

2. The number(s) 0 and 4 appear more than once.

3. The mode is the number that appears most often. The number 4 appears three times.

4. The mode is 4.

YOUR TURN!

Find the mode for the given set of data.

Ellis asked nine adults how many people they knew named Sarah.

$$3, 2, 6, 8, 6, 4, 6, 4, 1$$

1. Arrange the numbers in order from least to greatest.

_____

2. The number(s) _____ appear more than once.

3. The mode is the number that _____ _____. The number _____ appears _____ times.

4. The mode is _____.

Example 2

Find the median for the given set of data.

Romana recorded the ages of her cousins at a family reunion.

$$3, 12, 4, 2, 14, 5, 7, 12, 18, 16, 1$$

1. Arrange the numbers in order from least to greatest.

$$1, 2, 3, 4, 5, 7, 12, 12, 14, 16, 18$$

2. There are 11 numbers in the list. The sixth number, or the number in the middle, is the median.

3. The sixth number is 7.

$$\overbrace{1, 2, 3, 4, 5,}^{\text{5 terms}} \textcircled{7} \overbrace{12, 12, 14, 16, 18}^{\text{5 terms}}$$

YOUR TURN!

Find the median for the given set of data.

Wei-Ling recorded the number of coins she collected each day.

$$2, 10, 4, 3, 14, 5, 4, 12, 15,$$

1. Arrange the numbers in order from _____ to _____.

_____

2. There are _____ numbers in the list. The _____ number is the median.

3. The _____ number is ___.

Example 3

Find the range for the given set of data.

Hesutu asked nine adults how many miles they live from their work place.

14, 1, 8, 3, 11, 3, 21, 9, 17

1. Arrange the numbers in order from least to greatest.

    1, 3, 3, 8, 9, 11, 14, 17, 21

2. The least number is 1.
   The greatest number is 21.

3. Subtract the greatest number and the least number.

    21 − 1 = 20

4. The range is 20.

YOUR TURN!

Find the range for the given set of data.

Tariq asked nine adults how many candles they have in their homes.

4, 8, 8, 22, 31, 3, 21, 17, 15

1. Arrange the numbers in order from

    _____ to _____.

    _____

2. The _____ number is _____.

   The _____ number is _____.

3. Subtract the _____ number and

   the _____ number.

    _____ − _____ = _____

4. The range is _____.

Who is Correct?

Ava asked nine girls at what age they had their ears pierced. What is the mode of the data set?

7, 8, 4, 15, 15, 7, 3, 22, 5

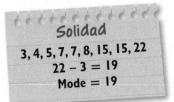

Solidad

3, 4, 5, 7, 7, 8, 15, 15, 22
22 − 3 = 19
Mode = 19

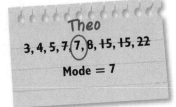

Theo

3, 4, 5, 7, (7), 8, 15, 15, 22
Mode = 7

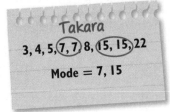

Takara

3, 4, 5, (7, 7), 8, (15, 15), 22
Mode = 7, 15

Circle the correct answer(s). Cross out incorrect answer(s).

 Guided Practice

Find the mode for each given set of data.

1 1, 7, 8, 1, 9, 11, 2

_____

2 5, 4, 13, 9, 26, 14, 21, 33, 9

_____

GO ON

Step by Step Practice

Find the median for the given set of data.

3 Jesse asked seven students in his class how many phone numbers they could name from memory.

$$5, 4, 13, 5, 9, 26, 21$$

Step 1 Arrange the data values in order from _____

to _____.

_____

Step 2 There are _____ numbers in the list. The _____ number, or the number in the middle, is the median.

Step 3 The _____ number is _____.

Find the median for each given set of data.

4 Grant asked seven adults how many plants they have in their home.

$$1, 7, 8, 1, 9, 11, 2$$

Arrange the numbers in order. _____, _____, _____, _____, _____, _____, _____

The median is _____.

5 Justin asked seven pet stores how many tropical fish they sold last week.

$$6, 7, 4, 5, 6, 13, 23$$

Arrange the numbers in order. _____, _____, _____,

_____, _____, _____, _____.

The median is _____.

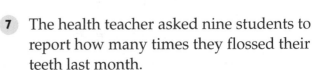

6 Kyle asked seven adults how many dollars they earned per hour.

$$18, 22, 20, 25, 31, 41, 20$$

The median is _____.

7 The health teacher asked nine students to report how many times they flossed their teeth last month.

$$9, 4, 13, 5, 26, 14, 21, 33, 9$$

The median is _____.

Find the range for each given set of data.

8 1, 7, 8, 1, 9, 11, 2

The range of the number of plants in Exercise 4 is

_____ − _____ = _____.

9 6, 7, 4, 5, 6, 13, 23

The range of the number of tropical fish sold in Exercise 5 is

_____ − _____ = _____.

10 18, 22, 20, 25, 31, 41, 20

The range in the dollars earned per hour in Exercise 6 is _____.

11 9, 4, 13, 5, 26, 14, 21, 33, 9

The range in the number of times students flossed in Exercise 7 is _____.

Step by Step **Problem-Solving Practice**

Solve.

12 **HEIGHT** Each member of the girls' volleyball team at Westminster Middle School measured their heights in inches and recorded the results. Find the mode, median, and range of the heights of the girls on the team.

64, 70, 65, 64, 65, 71, 65, 62, 63, 68, 66

Understand   Read the problem. Write what you know.

There are _____ numbers in the data set.

I want to find the _____, _____, and _____.

Plan   Pick a strategy. One strategy is to look for a pattern.

Solve   Arrange the numbers in order from least to greatest.

The _____, the number that appears most often,

is _____.

The _____, the number in the exact middle,

is _____.

The _____, the difference between the greatest number and the least number, is

_____ − _____ = _____.

Check   Reverse the order of the numbers and list them from greatest to least. The mode, range, and median should remain the same.

Problem-Solving Strategies
☐ Use a model.
☐ Use logical reasoning.
☐ Solve a simpler problem.
☐ Work backward.
☑ Look for a pattern.

GO ON

13 **CARNIVAL** The middle school had a carnival. Each school club had a booth that charged tickets for each activity. The data set shows the number of tickets each booth collected at the end of the day. Find the mode, median, and range of the tickets collected. Check off each step.

234, 196, 145, 254, 196, 223, 176, 155, 231

_____ Understand: I underlined key words.

_____ Plan: To solve the problem, I will _____.

_____ Solve: The answer is _____.

_____ Check: I checked my answer by _____.

14 **Reflect** How would adding the numbers 231 and 256 to the set of data below affect the median?

234, 196, 145, 254, 196, 223, 176, 155, 231

_____

_____

_____

Skills, Concepts, and Problem Solving

Find the mode for each given set of data.

15 4, 6, 7, 9, 7

_____

16 32, 53, 23, 24, 31, 44, 23

_____

17 6, 4, 7, 5, 19, 14, 19, 13, 8

_____

18 7, 6, 3, 6, 4, 8, 9, 4, 6, 4, 7

_____

Find the median for each given set of data.

19 Victoria asked seven adults how many dollars they spent in banking fees last year.

8, 6, 4, 8, 9, 3, 11

Arrange the numbers in order. _____, _____, _____, _____,

_____, _____, _____

The median is _____.

Find the median for each given set of data.

20 Gabriella kept a record of her team's score for the last five games.

52, 43, 26, 34, 29

The median is _____.

21 Melisa asked seven students the ages of their oldest living relative.

65, 77, 87, 91, 96, 77, 73

The median is _____.

Find the range for each given set of data.

22 16, 33, 17, 18, 22, 17, 21

_____ − _____ = _____

The range is _____.

23 8, 6, 4, 8, 9, 3, 11

_____ − _____ = _____

The range is _____.

24 52, 43, 26, 34, 29

The range is _____.

25 65, 77, 87, 91, 96, 77, 73

The range is _____.

Solve.

26 **MOVIES** A Hollywood movie company reported on the length, in minutes, of the last 13 movies they produced. The results are shown below. Find the mode, median, and range of the length of the movies.

83, 90, 121, 182, 122, 90, 97, 93, 122, 102, 99, 101, 92

_____

_____

27 **VIDEO GAMES** The newspaper surveyed middle school students and asked them how many video games they had at home. The results are shown below. Find the mode, median, and range of the data set.

11, 3, 2, 5, 13, 8, 9, 11, 5, 12, 4, 8, 6, 5, 7

_____

_____

GO ON

Vocabulary Check   **Write the vocabulary word that completes each sentence.**

28 The _____ is the value that falls exactly in the middle of a set of data.

29 The number that appears the most often in a data set is called the

_____.

30 The difference between the greatest number and the least number

in a data set is called _____.

31 **Writing in Math**   The data set below has an even number of data. How could you find the median, or the number in the middle?

2, 4, 8, 10, 12, 17

_____

_____

_____

 **Spiral Review**

32 Create a Venn diagram to sort the numbers. Classify them as factors of 24 or as prime numbers. (Lesson 7-1, p. 274)

2, 3, 4, 7, 8, 11, 12, 17, 20

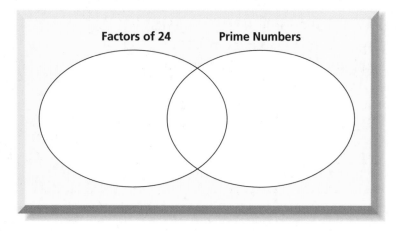

Factors of 24   Prime Numbers

STOP

Mean

KEY Concept

The **mean** is the sum of the numbers in a set of data divided by the number of pieces of data. Here are the award charts of four teams:

| 6 | 1 | 4 | 9 |

There are twenty ribbons in all. There are four teams. Redistribute the ribbons so that each team has the same number of ribbons.

| 5 | 5 | 5 | 5 |

If you redistribute the ribbons equally among the four teams, each team will have five ribbons. The mean of the number of ribbons for each team is five.

VOCABULARY

average
the sum of two or more quantities divided by the number of quantities; the mean

mean
the sum of the numbers in a set of data divided by the number of pieces of data

measures of central tendency
numbers that are often used to describe the center of a set of data; these measures include the mean, median, and mode

outlier
a value that is much higher or much lower than the other values of a set of data

A set of data may contain an **outlier**, which is a value that is much higher or lower than the other values. Outliers can greatly affect the mean. Sometimes an outlier is dropped from the set of data in order to describe the data better.

Example 1

Find the mean of 4, 9, 3, and 8.

1. Find the sum of the data. **4 + 9 + 3 + 8 = 24**

2. Count the items in the data set. There are **4** items in the data set.

3. Divide the sum by the number of items in the data set. **24 ÷ 4 = 6**

4. The mean of the data set is **6**.

GO ON

YOUR TURN!

Find the mean of 7, 2, 5, and 6.

1. Find the sum. _____ + _____ + _____ + _____ = _____

2. Count the items in the data set. There are _____ items in the data set.

3. Divide. _____ ÷ _____ = _____

4. The mean of the data set is _____.

Example 2

Find the mean of 5, 8, 4, and 1. Convert the remainder into a fraction or a decimal.

1. Find the sum of the data. $5 + 8 + 4 + 1 = 18$

2. Count the items in the data set.
   There are 4 items in the data set.

3. Divide the sum by the number $\dfrac{4 \text{ R2}}{4\overline{)18}}$
   of items in the data set.

4. Convert the remainder into a fraction $4 \text{ R2} = 4\dfrac{2}{4} = 4\dfrac{1}{2}$ or 4.5
   or a decimal.

5. The mean of the data set is 4.5 or $4\dfrac{1}{2}$.

YOUR TURN!

Find the mean of 3, 6, 1, and 4. Convert the remainder into a fraction or a decimal.

1. Find the sum. _____ + _____ + _____ + _____ = _____

2. Count the items in the data set.

   There are _____ items in the data set.

3. Divide. _____ ÷ _____ = _____

4. Convert the remainder into a fraction or a _____.

5. The mean of the data set is _____.

Example 3

The mean of three numbers is 5. Two of the numbers are 7 and 5.
Find the missing number.

1. Find the total value of 3 numbers with a mean of 5. $3 \cdot 5 = 15$

2. Find the sum of the given numbers. $7 + 5 = 12$

3. Subtract the sum of the two numbers from the total sum. $15 - 12 = 3$

4. The missing number is 3.

YOUR TURN!

The mean of three numbers is 4. Two of the numbers are 1 and 5.
Find the missing number.

1. Find the total value for 3 numbers with a mean of 4.

_____ $\cdot$ _____ $=$ _____

2. Find the sum of the given numbers.

_____ $+$ _____ $=$ _____

3. Subtract the sum of the two numbers from the total sum.

_____ $-$ _____ $=$ _____

4. The number missing from the data set is _____.

Who is Correct?

Find the mean of 5, 6, 5, 1, and 13.

Greg
The mean is 5 because it occurs twice in the data set.

Dwayne
The mean is 6.
$5 + 6 + 5 + 1 + 13 = 30$
$30 \div 5 = 6$

Greta
The mean is 11, because the least number is 1 and the greatest number is 12.

Circle correct answer(s). Cross out incorrect answer(s).

▶ Guided Practice

Find the mean of the data set. Convert the remainder into a fraction or a decimal.

1 9, 12, 7, 8

_____ + _____ + _____ + _____ = _____

_____ ÷ _____ = _____

The mean is _____.

2 7, 15, 12, 18

_____ + _____ + _____ + _____ = _____

_____ ÷ _____ = _____

The mean is _____.

3 10, 10, 10, 3, 5

_____ + _____ + _____ + _____ +

_____ = _____

_____ ÷ _____ = _____

The mean is _____.

4 4, 8, 5, 7, 8

_____ + _____ + _____ + _____ +

_____ = _____

_____ ÷ _____ = _____

The mean is _____.

Step by Step Practice

5 The mean of four numbers is 7. Three of the numbers are 9, 2, and 12. Find the missing number.

Step 1 Find the total value of 4 numbers with a mean of 7.

_____ • _____ = _____

Step 2 Find the sum of the given numbers.

_____ + _____ + _____ = _____

Step 3 Subtract the sum of the numbers from the total sum.

_____ − _____ = _____

Step 4 This missing number is _____.

Find one missing number from a data set when the mean is given.

6 Mean: 10   Data set: 8, 18, 6, _____

7 Mean: 3   Data set: 1, 4, 3, _____

8 Mean: 9   Data set: 4, 13, 9, _____

9 Mean: 7   Data set: 7, 8, 9, _____

Step by Step Problem-Solving Practice

Solve.

10 **NAMES** What is the mean number of letters in these students' names? CASSUNDRA, CARLY, KENT, and CARLOS.

Understand	Read the problem. Write what you know.
	There are _____ students.
	The number of letters in each name is _____,
	_____, _____, and _____.
Plan	Pick a strategy. One strategy is to act it out.
Solve	Use counters to represent the letters in each student's name. Find the _____ of the letters by counting the counters.
	Divide the counters by the number of _____.
	_____ ÷ _____ = _____
	The mean number of letters is _____.
Check	Use inverse operations and work backward.

11 **BASKETBALL** Five players on the girls' basketball team scored the following number of points in the first half of a game: 12, 4, 0, 8, and 5. What was the mean of the points scored? Check off each step.

_____ Understand: I underlined key words.

_____ Plan: To solve the problem, I will _____.

_____ Solve: The answer is _____.

_____ Check: I checked my answer by _____.

12 **TEST SCORES** Rochelle recorded the number of minutes she exercised over the last seven days. What is the mean of this data set?

64, 56, 60, 56, 68, 60, 63

_____

_____

13 **Reflect** Otis earned scores of 97, 93, 92, and 84 on his last four math tests. On the next test Otis, earned a score of 61. How does this outlier affect his mean score?

_____

_____

▶ Skills, Concepts, and Problem Solving

Find the mean of the data set.

14 97, 82, 71, 109, 91

_____

15 14, 26, 21, 29, 12

_____

16 4.6, 4.0, 3.6, 3.8, 4.9, 5.2, 6.1

_____

17 5.9, 6.1, 6.0, 5.9, 6.1

_____

Find the mean of the data set. Convert the remainder into a fraction or a decimal.

18 9, 4, 1, 8

_____

19 15, 7, 10, 6

_____

20 8, 12, 5, 6, 10

_____

21 11, 7, 6, 10, 20

_____

Find one missing number from a data set when the mean is given.

22 Mean: 10   Data set: 12, 15, _____

23 Mean: 8   Data set: 4, 12, 7, _____

24 Mean: 7   Data set: 11, 2, 8, _____

25 Mean: 11   Data set: 11.6, 12.5, 8.1, _____

26 The mean amount of money Jolene saved over the last six months was $52. She forgot to record the amount that she saved in April. What is the missing amount in the table at the right?

_____

Jolene's Savings Account	
Month	Dollars Saved ($)
March	46
April	
May	52
June	64
July	50
August	56

Solve.

27 **EXPERIMENT** A group of students grabbed a handful of marbles for a scientific experiment. Stephanie picked up 5 marbles, Hector picked up 8, Drew picked up 7 marbles, and Heddy picked up 4. What was the mean number of marbles the students grabbed?

_____

28 **QUIZ SCORES** Reyna scored 7, 8, 10, 6, and 8 on five weekly quizzes. What is the mean of her scores?

_____

Vocabulary Check   **Write the vocabulary word that completes each sentence.**

29 The sum of the numbers in a set of data divided by the number of

pieces of data is called the _____.

30 The _____ are numbers that are often used to describe the center of a set of data. This includes the mean, median, and mode.

31 **Writing in Math**   Is it necessary to order the data from least to greatest before you find the mean of the data? Explain your answer.

_____

_____

_____

 Spiral Review

Find the median and the range for each given set of data. (Lesson 7-2, p. 281)

32 Mariska recorded the ages of seven monkeys at the zoo.

8, 22, 15, 25, 36, 41, 30

The median is _____.
The range is _____.

33 The track coach asked nine students to report how many miles they ran last week.

9, 13, 15, 10, 16, 14, 21, 19, 9

The median is _____.
The range is _____.

STOP

Find the mode for each given set of data.

1 14, 26, 57, 39, 57

_____

2 26, 24, 27, 25, 39, 34, 39, 33, 28

_____

Find the median and the range for each given set of data.

3 Sarita asked seven adults how many cars they have owned in their lifetime.

1, 5, 2, 3, 7, 2, 4

Arrange the numbers in order. _____, _____, _____, _____, _____, _____, _____

The median is _____.
The range is _____.

Find the mean of the data set. Convert the remainder into a fraction or a decimal.

4 12, 5, 7, 8 _____

5 18, 12, 15, 9, 10 _____

6 **ANIMALS** Mr. Thomson listed these animals on the board: dog, lion, eagle, cat, elephant, and horse. He asked his class to sort these animals as pets or wild animals. Create a Venn diagram to sort these animals.

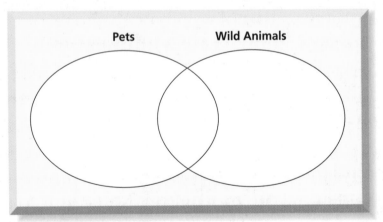

7 **SERVICE** Marco recorded the mean number of volunteers from the Nicolas Corporation for this year's charity events as 70 volunteers. However, he forgot to record the number of volunteers at the Children's Day event. Use the mean and the remaining data in the table to find the missing number.

Nicolas Corporation Charity Events	
Event	Number of Volunteers
Children's Day	
Health Fair	72
Golf Tournament	68
Jazz in the Park	52
Blankets for Babies	97

_____

Interpret Bar Graphs

KEY Concept

A **bar graph** is commonly used to compare categories of data. The graph has several important features, such as the title, categories, and a scale.

The categories are commonly set on the **horizontal axis** of the graph. The categories used in this graph are "Grade 5," "Grade 6," "Grade 7," and "Grade 8."

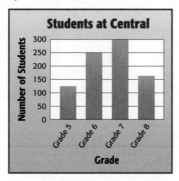

The **vertical axis** of the graph shows the scale of the numerical values. **Scales** use intervals to include data from larger value sets. Since the number of students at each grade ranges from 125 to 300 students, the **interval** used on the scale is 50.

Each bar represents a different grade. The bar height shows the number of students who are in each grade level.

The double-bar graph below shows the same set of data, but provides extra information about the number of boys and the number of girls in each grade level.

The interval on the scale has changed because the range of numbers is different.

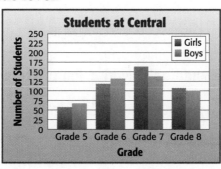

Copyright © Glencoe/McGraw-Hill, a division of The McGraw-Hill Companies, Inc.

VOCABULARY

bar graph
a graph using bars to compare quantities; the height or length of each bar represents a designated number

horizontal axis
the axis on which the categories or values are shown in a bar and line graph

interval
the difference between successive values on a scale

scale
the set of all possible values in a given measurement, including the least and greatest numbers in the set, separated by the intervals used

vertical axis
the axis on which the scale and interval are shown in a bar or line graph

The graphs above are vertical bar graphs. Horizontal bar graphs label the scales along the horizontal axis and the categories along the vertical axis.

GO ON

Example 1

Use the bar graph "My Classmates' Favorite Pants" to compare data.

How many more students prefer sweats than khakis?

1. How many students chose sweats as their favorite pants? **22**

2. How many students chose khakis as their favorite pants? **12**

3. To find how many more chose sweats than khakis, subtract.

   $22 - 12 = 10$

4. There are 10 more students who prefer sweats than khakis.

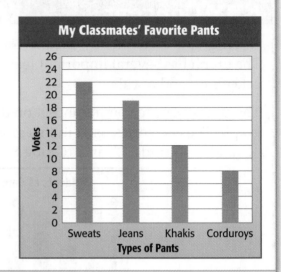

YOUR TURN!

Use the bar graph "My Classmates' Favorite Pants" to compare data.

How many students prefer jeans or corduroys?

1. How many students chose jeans as their favorite pants? _____

2. How many students chose corduroys as their favorite pants? _____

3. To find how many chose jeans or corduroys, _____.

   _____ + _____ = _____

4. There are _____ students who prefer jeans or corduroys.

Example 2

Use the double-bar graph "My Classmates' Favorite Pants" to compare data.

How many girls prefer jeans or corduroys?

1. How many girls chose jeans as their favorite pants? **8**

2. How many girls chose corduroys as their favorite pants? **3**

3. To find how many girls chose jeans or corduroys, add.

   $8 + 3 = 11$

4. There are 11 girls who prefer jeans or corduroys.

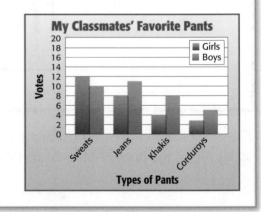

YOUR TURN!

Use the double-bar graph "My Classmates' Favorite Pants" on page 298 to compare data.

How many more boys than girls prefer khakis?

1. How many boys chose khakis as their favorite pants? _____

2. How many girls chose khakis as their favorite pants? _____

3. To find how many more boys chose khakis than girls, _____.

    _____ − _____ = _____

4. There are _____ more boys than girls who prefer khakis.

Who is Correct?

Use the double-bar graph "My Classmates' Favorite Pants" on page 298 to compare data.

Compare the number of boys who chose jeans, and the number of boys who chose sweats as their favorite pants.

Mallory
12 − 8 = 4
Four more boys chose
jeans than sweats.

Tadeo
11 − 10 = 1
One more boy chose
jeans than sweats.

Kei
11 − 8 = 3
Three more boys chose
jeans than sweats.

Circle correct answer(s). Cross out incorrect answer(s).

 Guided Practice

Use the bar graph "Mary's Guitar Practice" to compare data.

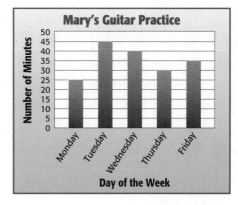

1 What does the scale of the bar graph represent?

_____

2 What are the categories?

_____

3 What interval is used for the scale? _____

4 What does the height of each bar represent?

_____

GO ON

Use the bar graph "Mary's Guitar Practice" on page 299 to compare data.

5 How many more minutes did Mary practice on Tuesday than on Thursday?

Step 1 How many minutes did Mary practice on Tuesday?

_____

Step 2 How many minutes did Mary practice on Thursday?

_____

Step 3 To find how many more minutes she practiced on Tuesday than on Thursday, _____.

_____ – _____ = _____

Step 4 Mary practiced _____ minutes more on _____.

Use the bar graph "Favorite Type of TV Show" to compare data.

6 How many students chose horror shows as their favorite?

_____

7 What number of students chose action shows as their favorite?

_____

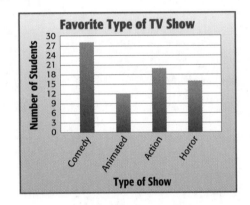

8 How many students chose comedy or animated shows as their favorite?

_____

9 Compare the number of students who chose comedy shows to the number who chose horror shows.

_____

10 Find the total number of students represented on the graph.

_____

11 Which type of TV show is preferred the most?

_____

Step by Step Problem-Solving Practice

Solve.

Problem-Solving Strategies
- ☐ Draw a diagram.
- ☐ Use logical reasoning.
- ☑ Solve a simpler problem.
- ☐ Work backward.
- ☐ Look for a pattern.

12 **CAFETERIA** The cafeteria manager wanted to compare the favorite meals of the students at Madison Middle School. How many girls prefer pizza or spaghetti compared to the number of girls who prefer chicken nuggets or tacos?

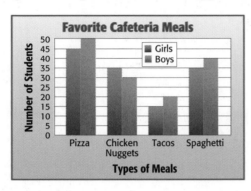

Understand Read the problem and review the graph. Write what you know.

There are _____ girls who prefer pizza.

There are _____ girls who prefer chicken nuggets.

There are _____ girls who prefer tacos.

There are _____ girls who prefer spaghetti.

Plan Pick a strategy. One strategy is to solve a simpler problem.

Solve How many girls prefer pizza or spaghetti?

_____ + _____ = _____

How many girls prefer chicken nuggets or tacos?

_____ + _____ = _____

What is the difference between these two groups?

_____ − _____ = _____

There are _____ girls who prefer pizza or spaghetti compared to chicken nuggets or tacos.

Check Work backward and use inverse operations. Use addition to check subtraction, and subtraction to check addition.

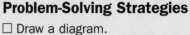

Use the double-bar graph "Favorite Cafeteria Meals" on page 301 to complete Exercises 13 and 14.

13 FOOD The cafeteria manager wanted to compare the favorite meals of boys and girls. How many girls prefer pizza or spaghetti compared to the number of boys who prefer pizza or spaghetti? Check off each step.

_____ Understand: I underlined key words.

_____ Plan: I will solve this problem by _____.

_____ Solve: The answer is _____.

_____ Check: To check my answer I will _____.

14 PIZZA How many more students (boys and girls) preferred pizza compared to tacos?

_____

_____

15 Reflect Other than comparing boys and girls, what other two groups could be compared using a double-bar graph?

_____

_____

▶ Skills, Concepts, and Problem Solving

Use the bar graph "Favorite School Subject" to compare data.

16 How many students chose math as their favorite subject?

_____

17 How many students chose history as their favorite subject?

_____

18 How many students chose reading or creative writing as their favorite subject?

_____

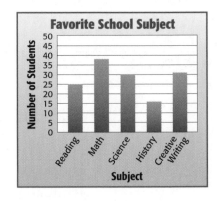

Use the bar graph "Favorite School Subject" on page 302 to compare data.

19 How many more students chose science than history as their favorite subject?

_____

20 How many students are represented on the graph?

_____

21 Which subject is preferred the least?

_____

Use the double-bar graph "Favorite Sport" to complete Exercises 22 and 23.

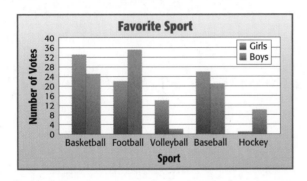

22 SPORTS   The physical education teacher wanted to compare the favorite sports of boys and girls. How many girls prefer basketball or baseball compared to the number of boys who prefer basketball or baseball?

_____

_____

_____

_____

23 FOOTBALL   How many more students (boys and girls) prefer football compared to hockey?

_____

_____

_____

GO ON

Vocabulary Check   **Write the vocabulary word that completes each sentence.**

24 A _____ is a graph using bars to compare quantities.

25 The _____ is commonly the axis on which the scale and interval are shown in a bar or line graph.

26 **Writing in Math**   Suppose you are creating a bar graph. Choose a scale for the following numbers. Include information about the interval you would use. Explain your answer.

56, 97, 125, 79, 205, 152

_____

_____

_____

 Spiral Review

Find one missing number from a data set when the mean is given.
(Lesson 7-3, p. 289)

27 Mean: 16

Data set: 18, 21, 19, _____

28 Mean: 10

Data set: 7, 16, 9, _____

29 Mean: 19

Data set: 24, 23, 27, _____

30 Mean: 46

Data set: 47, 45, 46, _____

31 Create a Venn diagram to sort the numbers. Classify them as negative integers or odd numbers. (Lesson 7-1, p. 274)

−4, 5, −7, 9, −1, 17, 24, −18

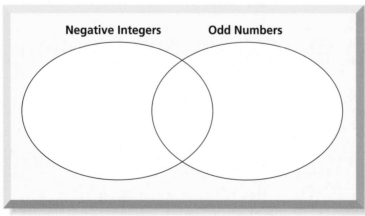

Create Bar Graphs

KEY Concept

A **bar graph** can be created using information in a table. The final graph must show several important features, such as the title, categories, and a **scale**.

Pets Owned	
Pet	Number
Cat	10
Dog	12
Fish	5
Bird	2
Lizard	1
None	2

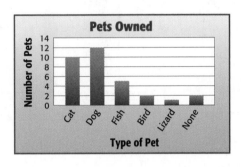

Notice that the title and the categories in the graph are similar to those shown in the table. The scale of the graph often starts at zero and extends to the largest value in the data set. The **interval** of the scale is determined by the range of the data. This graph uses an interval of 2.

A double-bar graph can also be created using the information in a table. This double-bar graph compares pet ownership of boys and girls.

Pets Owned		
Pet	Boys	Girls
Cat	2	8
Dog	6	6
Fish	3	2
Bird	0	2
Lizard	1	0
None	1	1

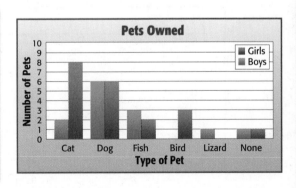

Double-bar graphs can compare categories other than boys and girls, such as two grade levels, adults and children, or other groups.

VOCABULARY

bar graph
a graph using bars to compare quantities; the height or length of each bar represents a designated number

horizontal axis
the axis on which the categories or values are shown in a bar and line graph

interval
the difference between successive values on a scale

scale
the set of all possible values in a given measurement, including the least and greatest numbers in the set, separated by the intervals used

vertical axis
the axis on which the scale and interval are shown in a bar or line graph

Example 1

Use the data in the table to create a bar graph.

The table shows the number of points scored last week by the top four players on the Carlson Middle School boys' basketball team.

Number of Points Scored	
Name	Points
Tai	24
Jordan	12
Gregorio	17
Chandler	14

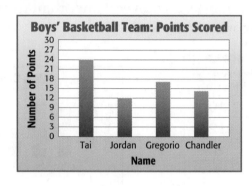

1. Write the title.

2. Label the horizontal and vertical axes.

3. Choose the interval and complete the scale.

4. Draw the bars to represent the number of points.

YOUR TURN!

Use the data in the table to create a bar graph.

The table shows the number of points scored last week by the top four players on the Giovanni Middle School girls' basketball team.

Number of Points Scored	
Name	Points
Barb	16
Elisa	18
Kanisha	10
Farah	6

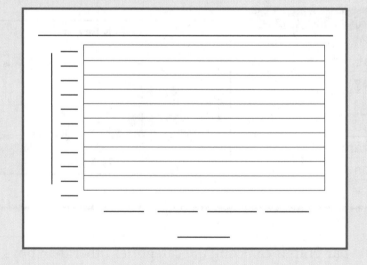

1. Write the title.

2. Label the horizontal and vertical axes.

3. Choose the interval and complete the scale.

4. Draw the bars to represent the number of points.

Example 2

Use the data in the table to create a double-bar graph.

The table shows the number of items collected last week by Room 101 and Room 102 in the Lincoln Middle School charity clothing drive.

1. Write the title.

2. Label the horizontal and vertical axes.

3. Choose the interval and complete the scale.

4. Draw the bars to represent the number of items.

5. Make a key to show the data for Room 101 and Room 102.

Clothing Drive Items

Type of Clothing	Number of Items	
	Room 101	Room 102
Shirts	21	18
Pants	17	24
Coats	3	7
Pairs of shoes	16	11

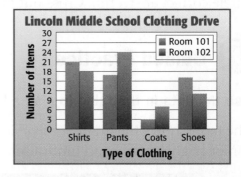

YOUR TURN!

Use the data in the table to create a double-bar graph.

The table shows the number of items collected last week by Grade 7 and Grade 8 in the Lincoln Middle School charity clothing drive.

1. Write the title.

2. Label the horizontal and vertical axes.

3. Choose the interval and complete the scale.

4. Draw the bars to represent the number of items.

5. Make a key to show the data for Grade 7 and Grade 8.

Clothing Drive Items

Type of Clothing	Number of Items	
	Grade 7	Grade 8
Shirts	28	39
Pants	32	40
Coats	15	10
Pairs of shoes	19	27

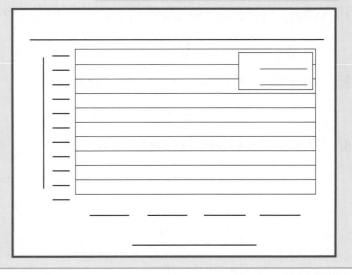

Who is Correct?

Use the values below to choose an interval for the scale of a bar graph.

19, 20, 16, 11, 7

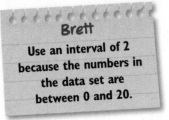

Brett

Use an interval of 2 because the numbers in the data set are between 0 and 20.

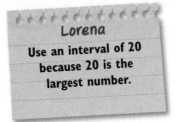

Lorena

Use an interval of 20 because 20 is the largest number.

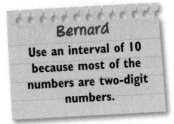

Bernard

Use an interval of 10 because most of the numbers are two-digit numbers.

Circle correct answer(s). Cross out incorrect answer(s).

 Guided Practice

Use the data in the table to plan a bar graph.

The table shows the number of vehicles that were washed at the Jefferson Junior High School band fundraiser.

Jefferson JHS Band Car Wash	
Vehicle	Number Washed
Van	15
SUV	19
Sedan	11
Sports Car	3

1 What is a good title for the graph?

_____

2 What labels could be used for the *x*- and *y*-axes?

_____

3 What interval could be used for the scale?

_____

4 What will the height of each bar represent?

_____

Step by Step Practice

Use the data in the table on page 308 to create a bar graph.

5 **Step 1** Write the title.

Step 2 Label the horizontal and vertical axes.

Step 3 Choose the interval and complete the scale.

Step 4 Draw the bars to represent the number of vehicles washed.

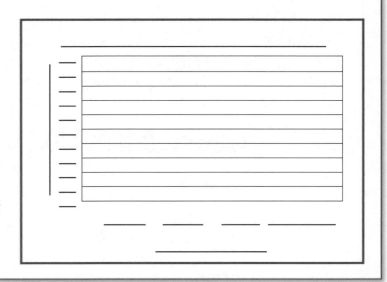

Use the data in each table to create a bar graph.

6 The table shows the number of instruments played in the Jefferson Junior High School orchestra.

Number of Instruments Played	
Instrument	Number
Woodwind	15
Brass	8
Percussion	7
String	19

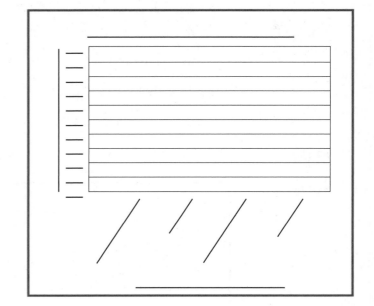

7 The table shows the number of singers in the Jefferson Junior High School choir.

Singers in Jefferson JHS Choir	
Type	Number
Soprano	15
Alto	12
Tenor	7
Bass	3

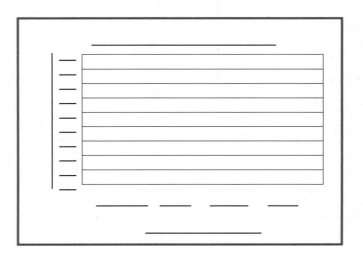

Step by Step Problem-Solving Practice

Problem-Solving Strategies
☑ Use a table.
☐ Use logical reasoning.
☐ Solve a simpler problem.
☐ Work backward.
☐ Look for a pattern.

Use the data in the table to create a double-bar graph.

8 ASSEMBLIES The vice principal wanted to compare the favorite assemblies of the students at Grant Junior High School using data from the table. She will present the graph at the next parent-teacher meeting.

Favorite Grant JHS Assemblies		
Type of Assembly	Number of Votes	
	Grade 7	Grade 8
School Spirit	74	57
Music and the Arts	12	31
Health Education	10	4
Community Service	24	33

Understand   Read the table. Write what you know.

The interval of the scale could be _____.

The key will show the data for

_____ and _____.

Plan   Pick a strategy. One strategy is to use a table.

Solve

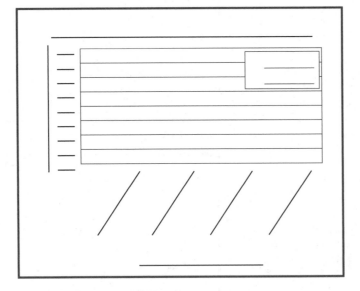

Check   Check the length of each bar on the graph to be certain that it matches the data in the table.

Use the data in the table to create a double-bar graph.

9 **FIELD TRIPS** The vice principal wanted to compare the favorite field trips of the boys and girls from the table. She will present a double-bar graph at the next parent-teacher meeting. Check off each step.

Favorite Field Trips		
Type of Field Trip	Number of Votes	
	Boys	Girls
Aquarium	63	68
Science Museum	19	24
Ballet	11	3
Community Theater	25	32

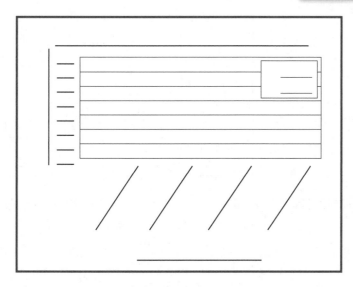

_____ **Understand: I underlined key words.**

_____ **Plan: I will solve this problem by** _____.

_____ **Solve: The answer is** _____.

_____ **Check: To check my answer I will** _____.

10 **Reflect** How would the graph in Exercise 9 be different if zero girls had chosen the ballet as their favorite field trip?

_____

_____

GO ON

Use the data in the table to create a bar graph.

11 The table shows the favorite types of music for Mr. Augayo's class.

Favorite Music of Mr. Augayo's Students	
Type	Number of Votes
Rock	5
Pop	17
R & B or Rap	7
Techno	1

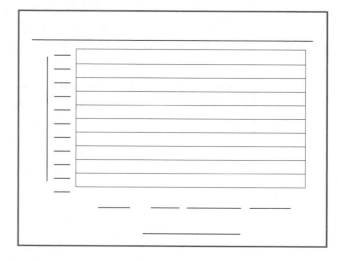

Use the data in the table to create a double-bar graph.

12 **FUND-RAISING**   The principal wants to make a double-bar graph to compare the number of dollars raised at each fund-raising event at Martin Middle School in 2008 and 2009. She will present the graph at the next parent-teacher meeting.

Martin Middle School Fund-raisers		
Event	Dollars Raised	
	2008	2009
Car Wash	225	175
Gift Wrap Sales	200	250
Walk-A-Thon	75	100
Pizza Night	125	150

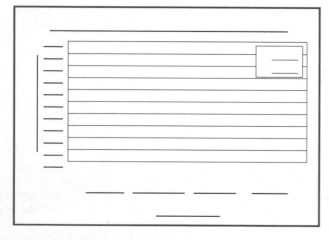

Vocabulary Check   **Write the vocabulary word that completes each sentence.**

13 The _____ is the difference between successive values on a scale.

14 The _____ is commonly the axis on which the categories or values are shown in a bar and line graph.

15 The _____ is the axis on which the scale and interval are shown in a bar or line graph.

16 **Writing in Math**   Suppose you are creating a bar graph. Choose a scale for the following numbers. Include information about the interval you would use. Explain your answer.

56, 97, 125, 79, 205, 152

_____

_____

_____

▶ Spiral Review

Use the bar graph created in Exercise 12 to compare data. (Lesson 7-4, p. 297)

17 How much money was raised by the Walk-A-Thon in 2008?

_____

18 How much more money was raised by the car wash in 2008 than 2009?

_____

19 How much money was raised in gift wrap sales in 2008 and 2009?

_____

20 In which year was the most money raised?

_____

STOP

Use the bar graph "Carnival Sales" to compare data.

1 How many people purchased a slice of pizza at the carnival?

_____

2 How many more people purchased a hamburger than purchased a hot dog?

_____

3 How many more people purchased a slice of pizza than hot dogs and hamburgers combined?

_____

4 How many items were sold at the carnival?

_____

Use the data in the table to create a double-bar graph.

5 **MUSIC** Summer needs to create a double-bar graph for her math class. She recorded the songs that her friends downloaded last week. She will share the graph with her friends at school.

Number of Songs Downloaded		
Name	**Number of Songs**	
	Rock	**Pop**
Katie	16	5
Kiyo	6	12
Jocelyn	5	15
Tavio	19	0

Interpret Line Graphs

KEY Concept

A **line graph** is used to show how a set of data changes over a period of time. Like a bar graph, it has features such as a title, categories, and a scale.

The scale shown on a line graph is commonly set on the **vertical axis**.

The **horizontal axis** of the graph commonly shows the period of time such as hours, years, or decades.

Since the number of dollars raised each week in the graph below varies from 250 to 1,200, the **interval** on the scale is 250.

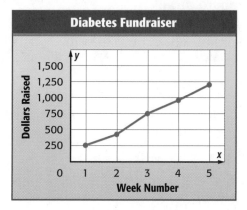

A double-line graph compares two sets of data. The lines of each set of data are a different color to make the graph easier to read. The line graph below compares the amount of money raised in 2008 and 2009.

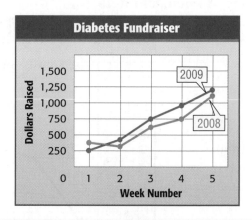

VOCABULARY

horizontal axis
the axis on which the categories or values are shown in a bar and line graph

interval
the difference between successive values on a scale

line graph
a graph used to show how a set of data changes over a period of time

scale
the set of all possible values in a given measurement, including the least and greatest numbers in the set, separated by the intervals used

vertical axis
the axis on which the scale and interval are shown in a bar or line graph

The points on a line graph are plotted like ordered pairs on a coordinate grid.

Example 1

Use the line graph "Max's Income" to compare data.

How much more does Max earn after 5 hours of
work compared to 4 hours of work?

1. How much did Max earn after 4 hours? $30

2. How much did Max earn after 5 hours? $37.50

3. To find how much more Max earned, subtract.

   $37.50 − $30.00 = $7.50

4. Max earned $7.50 more.

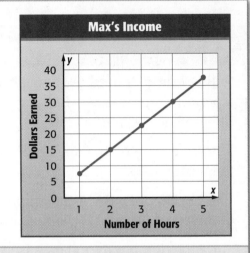

Max's Income

YOUR TURN!

Use the line graph "Max's Income" to compare data.

How much more does Max earn after 3 hours of work compared to
1 hour of work?

1. How much did Max earn after 3 hours? _____

2. How much did Max earn after 1 hour of work? _____

3. To find how much more he earned, _____.

   _____ − _____ = _____

4. Max earned _____ more.

Example 2

Use the double-line graph "School Spirit Sales" to compare data.

How much more was spent on T-shirts
than on buttons in September?

1. How much was spent on T-shirts in September?

   $600

2. How much was spent on buttons in September?

   $100

3. To find the difference in sales, subtract.

   $600 − $100 = $500

4. The amount spent on T-shirts was $500 more.

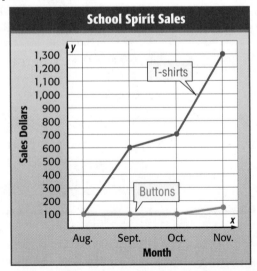

School Spirit Sales

YOUR TURN!

Use the double-line graph "School Spirit Sales" to compare data.

How much more was spent on T-shirts in November than in August?

1. How much was spent on T-shirts in November? _____

2. How much was spent on T-shirts in August? _____

3. To find how much more was spent, _____.

   _____ − _____ = _____

4. The amount spent on T-shirts in November was _____ more.

Who is Correct?

Use the double-line graph "School Spirit Sales" to compare data.

Compare the T-shirt sales and the button sales in September.

Andre
$100 − $100 = $0
There was no difference.

Michelle
$600 − $100 = $500
T-shirt sales were
$500 more.

Zacharias
$600 − $100 = $500
Button sales were
$500 less.

Circle correct answer(s). Cross out incorrect answer(s).

 Guided Practice

Use the line graph "Jason's Savings Account" to compare data.

1 What does the vertical axis of the bar graph represent?

_____

2 What is indicated on the horizontal axis?

_____

3 What does the dot on the horizontal axis indicate?

_____

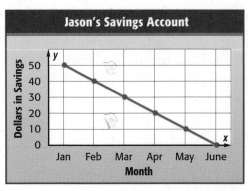

Use the line graph "Jason's Savings Account" on page 317 to compare data.

4　How much money does Jason take out of his savings during the month of February?

Step 1　The value of the account at the beginning of February is _____.

Step 2　The value of the account at the beginning of March is _____.

Step 3　To find the amount that was withdrawn, _____.

_____ − _____ = _____

Step 4　Jason withdrew _____ in February.

Use the line graph "Janelle's Heart Rate" to compare data.

5　What is Janelle's heart rate at the 15-minute mark?

_____

6　What is Janelle's heart rate at the 40-minute mark?

_____

7　Compare Janelle's heart rate at the 10-minute mark to her heart rate at the 30-minute mark.

_____

_____

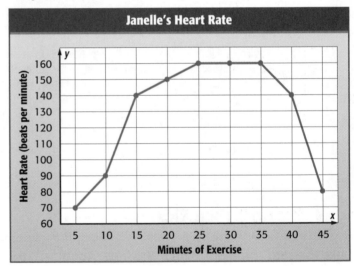

8　Compare Janelle's heart rate at the 5-minute mark to her heart rate at the 45-minute mark.

_____

_____

9　Describe the trend of the data. Does it increase, decrease, or stay the same over time?

_____

Step by Step Problem-Solving Practice

Use the double-line graph "Motor Vehicle Mileage" to compare data.

10 MILEAGE Jordan created a line graph comparing the yearly mileage of Mr. Lane's semi-trailer truck and his personal vehicle. Compare the average annual mileage of the truck to his personal vehicle in 2010.

Problem-Solving Strategies

☐ Draw a diagram.
☑ Use a graph.
☐ Solve a simpler problem.
☐ Work backward.
☐ Look for a pattern.

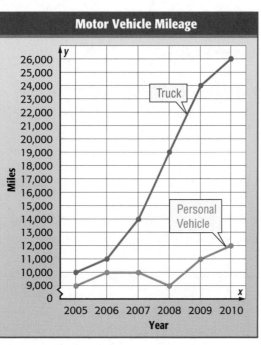

Understand Read the problem and review the graph. Write what you know.

The mileage for the truck is shown in _____.

The mileage for the personal vehicle is shown

in _____.

Plan Pick a strategy. One strategy is to use a graph.

Solve The truck drove _____ miles in 2010.

The personal vehicle drove _____ miles in 2010.

To compare the mileage, _____.

_____ − _____ = _____

Check Use inverse operations.

GO ON

Use the double-line graph "Motor Vehicle Mileage" on page 319 to compare data.

11 TRUCKS   The owner of Tampera Trucking Company wants to compare the mileage Mr. Lane's truck traveled in 2010 to 2005. Check off each step.

_____ **Understand: I underlined key words.**

_____ **Plan: I will solve this problem by** _____.

_____ **Solve: The answer is** _____.

_____ **Check: To check my answer I will** _____.

12 VEHICLES   Compare the annual mileage of Mr. Lane's personal vehicle in 2006 and 2008.

_____

_____

13 Reflect   What other interval could be used for the graph "Motor Vehicle Mileage?" How would this change the data?

_____

_____

_____

 Skills, Concepts, and Problem Solving

Use the line graph "Donations Collected by Amelia's Class" to compare data.

14 How much money did Amelia's class collect in September?

_____

15 How much money did Amelia's class collect in June?

_____

16 How much more was collected in March than in December?

_____

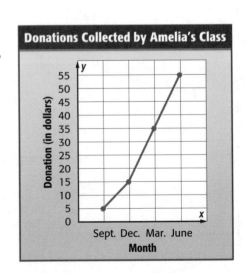

Donations Collected by Amelia's Class

Use the line graph "Donations Collected by Amelia's Class" on page 320 to compare data.

17 What is the interval of the scale?_____

18 How much money was collected in March?_____

19 Describe the trend of the graph. Do the donations increase, decrease, or show no change?

_____

Use the double-line graph "Theater Rehearsal Time" to complete Exercises 20–22.

20 **THEATER**   The theater teacher made a graph of the rehearsal time at Monterrey High School over the past few weeks. Compare the number of minutes actors rehearsed during Week 3 to the number of minutes the stage crew rehearsed during Week 3.

_____

_____

_____

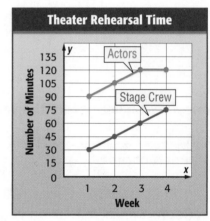

21 **STAGE CREW**   Compare the number of minutes the stage crew spent in rehearsal during Week 4 to the number of minutes the stage crew spent in rehearsal during Week 1.

_____

_____

_____

22 **PATTERNS**   If the trends continue, what would you expect the data to show for Week 5?

_____

_____

_____

Use the double-line graph "Concession Stand Sales" to complete Exercises 23 and 24.

23 **SALES** The concession stand manager made a graph to compare sales of bottled water and hot cocoa during the football season. Compare the sales of water and cocoa in October.

_____

_____

24 **PATTERNS** If the trends continue, what would you expect the data to show for November?

_____

_____

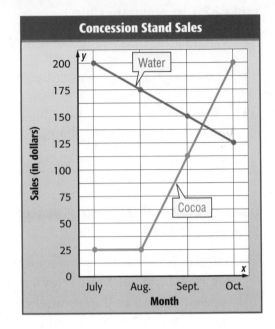

Concession Stand Sales

Vocabulary Check **Write the vocabulary word that completes the sentence.**

25 A _____ is used to show how a set of data changes over a period of time.

26 **Writing in Math** Give a real-world example to explain when line graphs are used. Explain your answer.

_____

_____

 Spiral Review

Use the data in the table to plan a bar graph. (Lesson 7-5, p. 304)

The table shows the number of athletes at the Johnson Middle School.

27 What is a good title for the graph?

_____

28 What are the two main categories?

_____

29 What interval could be used for the scale? _____

Johnson Middle School Athletes	
Sport	Number of Students
Soccer	21
Baseball	19
Football	30
Basketball	37

Create Line Graphs

KEY Concept

Information from a table can be used to create a **line graph**. The line graph below describes the number of miles Mr. Batista ran while he was training for a marathon. Notice that the title and the categories of the graph are similar to those shown in the table.

Marathon Training	
Month	Miles
January	5
February	12
March	18
April	21
May	25
June	30

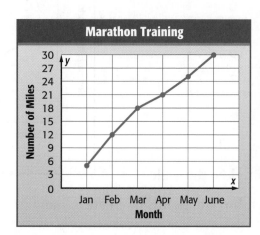

The scale of the graph often starts at zero and the **interval** of the scale is determined by the range of the data.

A double-line graph can also be created using the information in a table. The double-line graph below compares donations to dog and cat rescue.

Donations (in thousands)		
Year	Dogs	Cats
2004	$3	$2
2005	$6	$6
2006	$8	$10
2007	$13	$8
2008	$12	$13
2009	$14	$16

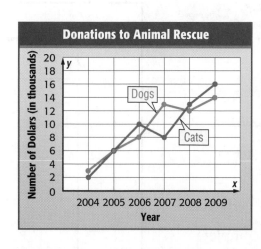

VOCABULARY

horizontal axis
the axis on which the categories or values are shown in a bar or line graph

interval
the difference between successive values on a scale

line graph
a graph used to show how a set of data changes over a period of time

scale
the set of all possible values in a given measurement, including the least and greatest numbers in the set, separated by the intervals used

vertical axis
the axis on which the scale and interval are shown in a bar or line graph

The lines in a double-line graph are often color-coded and labeled to indicate the categories.

Example 1

Use the data in the table to create a line graph.

The table shows the number of hours Lucas spent completing his homework over a four-week period.

Lucas's Homework Log	
Week of	Hours
April 8	5
April 15	4
April 22	2
April 29	7

1. Write the title.

2. Label the horizontal and vertical axes.

3. Choose the interval and complete the scale.

4. Plot each point and connect the data points to create a line.

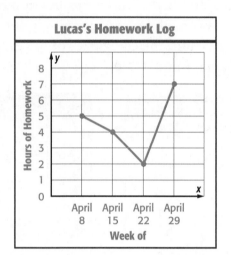

Lucas's Homework Log

YOUR TURN!

Use the data in the table to create a line graph.

The table shows the number of cars sold over a four-month period by Argosky Auto Mall.

Argosky Car Sales	
Month	Sales Dollars (in thousands)
April	27
May	32
June	19
July	12

1. Write the title.

2. Label the horizontal and vertical axes.

3. Choose the interval and complete the scale.

4. Plot each point and connect the data points to create a line.

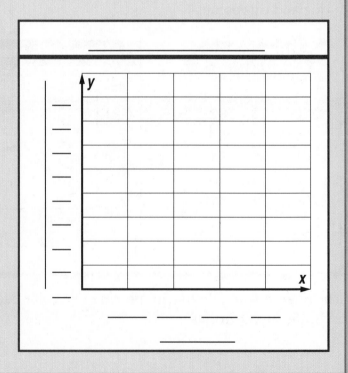

Example 2

Use the data in the table to create a double-line graph.

The table shows the values of two houses of the same size in Kansas and Florida over a period of four years.

Housing Values		
Year	Value (in thousands)	
	Kansas	Florida
2006	95	150
2007	100	200
2008	120	250
2009	115	230

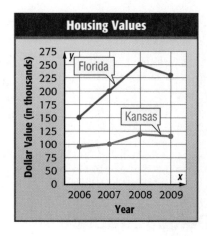

1. Write the title.

2. Label the horizontal and vertical axes.

3. Choose the interval and complete the scale.

4. Plot each point one line at a time.

5. Use a different color for each line and color-code each line's label.

YOUR TURN!

Use the data in the table to create a double-line graph.

The table shows the number of yards Amy and Isa walked in the Washington Middle School Walk-A-Thon.

Walk-A-Thon Results		
Number of Minutes	Yards Walked	
	Amy	Isa
10	14	19
20	30	36
30	44	54
40	54	66

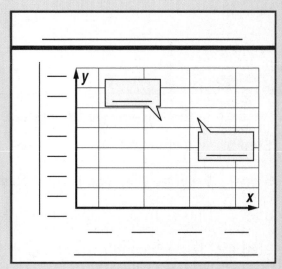

1. Write the title.

2. Label the horizontal and vertical axes.

3. Choose the interval and complete the scale.

4. Plot each point one line at a time.

5. Use a different color for each line and color-code each line's label.

GO ON

Who is Correct?

Use the line graph "Walk-A-Thon Results" on page 325 to compare the data.

Describe the trend of the data. Do the number of yards increase, decrease, or stay the same over time?

Alicia

The data remains the same because Amy and Isa walk at the same rate.

Vidia

The data shows an increase over time because Amy and Isa continue to walk.

Tomás

The data decreases because Amy and Isa slow down.

Circle correct answer(s). Cross out incorrect answer(s).

 Guided Practice

Use the data in the table to plan a line graph.

The table shows the number of staffing hours at Phillips Bakery in a four-hour time period.

1 What interval could be used for the scale?

_____

2 Describe the trend of the data. Do the number of staffing hours increase, decrease, or stay the same over time?

_____

Phillips Bakery	
Time	Staffing Hours
7 A.M.	15
8 A.M.	34
9 A.M.	46
10 A.M.	56

Step by Step Practice

Use the data in the table above to create a line graph.

3 **Step 1** Write the title.

Step 2 Label the horizontal and vertical axes.

Step 3 Choose the interval and complete the scale.

Step 4 Plot each point and connect the data points to create a line.

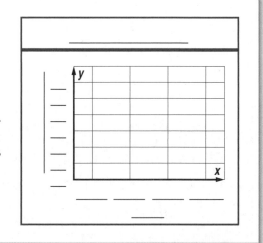

Step by Step Problem-Solving Practice

Use the data in the table to create a double-line graph.

(4) **FINANCIAL LITERACY** Mr. and Mrs. Cho have been saving money to purchase a new couch. The table shows how much money each person has in their savings account at the end of each month.

Problem-Solving Strategies
☐ Draw a diagram.
☐ Work backward.
☐ Solve a simpler problem.
☑ Use a table.
☐ Look for a pattern.

Cho Family Savings		
Month	**Number of Dollars ($)**	
	Mr. Cho	**Mrs. Cho**
January	740	625
February	870	780
March	420	420
April	675	725

Understand Read the table. Write what you know.

The interval of the scale could be _____.

The _____ will show the data for _____.

The _____ will show data for _____.

Plan Pick a strategy. One strategy is to use a table.

Solve

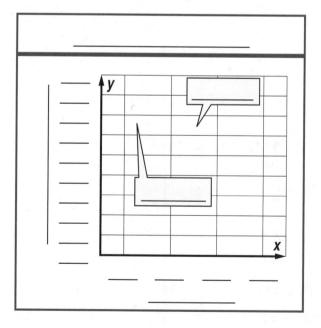

Check Check the color-coding on each line and compare the data on the graph to the data in the table.

GO ON

Use the data in the table to create a double-line graph.

5 **HOMEWORK** The PTA president will present a double-line graph showing the average amount of study time per month of the boys and girls at Huntsville Junior High School. Check off each step.

Study Time		
Year	Number of Hours	
	Boys	Girls
2006	175	150
2007	150	175
2008	200	225
2009	175	200

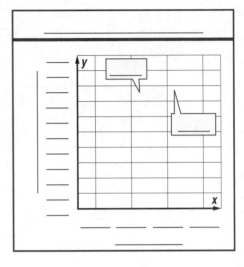

_____ Understand: I underlined key words.

_____ Plan: I will solve this problem by _____.

_____ Solve: The answer is _____.

_____ Check: To check my answer I will _____.

6 **Reflect** Look at the line graph for Exercise 5. Describe the trend of the data. Do the number of hours studying increase, decrease, or show no relationship?

_____

_____

 Skills, Concepts, and Problem Solving

Use the data in the table to create a line graph.

7 The table shows Marka's hourly wages over a four-year time period.

Marka's Hourly Wages	
Year	Hourly Wages
2006	7
2007	10
2008	10
2009	12

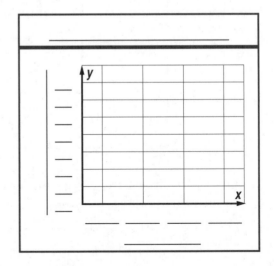

Use the data in the table to create a double-line graph.

8 **SALES** The table shows the sales dollars for PWR 4 and RX 71 cell phones sold at Kingston Electronics over a four-month period.

Kingston Electronics Sales		
Month	Cell Phone Model	
	PWR 4	RX 71
May	$125,000	$175,000
June	$150,000	$75,000
July	$175,000	$100,000
August	$225,000	$150,000

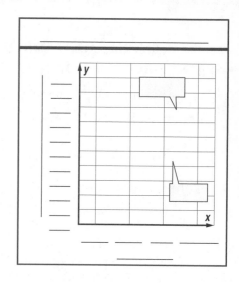

Vocabulary Check **Write the vocabulary word that completes each sentence.**

9 The _____ is the difference between successive values on a scale.

10 The _____ is commonly the axis on which the scale and interval are shown in a bar or line graph.

11 **Writing in Math** Compare a bar graph and a line graph. Explain the differences and similarities between the two graphs.

_____

_____

_____

 Spiral Review

Use the line graph created in Exercise 8 to compare data. (Lesson 7-6, p. 315)

12 What were the sales of RX 71 phones in August?

_____

13 How much higher were sales of PWR 4 phones than sales of RX 71 phones in June?

_____

14 Compare the sales of the PWR 4 and RX 71 phones. Which phone is more popular?

_____

15 Describe the trend of the data for the PWR 4 phone. Do the sales increase, decrease, or stay the same over time?

_____

Use the double-line graph "Ares Amusement Parks Ticket Sales" to compare data.

1   Describe the sales in 2007 for the Banchee Haunted House.

_____

2   Compare the sales of the Titan Tidal Pool in 2008 to 2006.

_____

3   Compare the sales of the Titan Tidal Pool to the Banchee Haunted House in 2008.

_____

4   Compare the sales of the Banchee Haunted House in 2007 to 2006.

_____

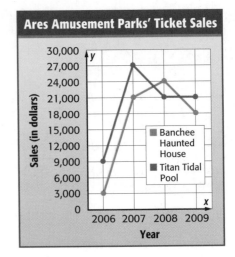

Use the data in the table to create a line graph.

5   **SALES**   Donavan needs to create a line graph for his boss. He recorded the number of cars that were sold at the dealership this week.

Graeter Deal Used Car Sales	
Day of the Week	Sales
Saturday	$19,000
Sunday	$17,000
Monday	$17,000
Tuesday	$15,000
Wednesday	$12,000

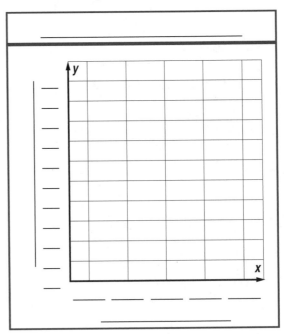

Vocabulary and Concept Check

average, *p. 289*

bar graph, *p. 297*

data, *p. 281*

horizontal axis, *p. 297*

interval, *p. 297*

line graph, *p. 315*

mean, *p. 289*

measures of central tendency, *p. 289*

median, *p. 281*

mode, *p. 281*

outlier, *p. 289*

range, *p. 281*

scale, *p. 297*

sort, *p. 274*

Venn diagram, *p. 274*

vertical axis, *p. 297*

Write the vocabulary word that completes each sentence.

1 A(n) _____ is a diagram that uses overlapping and separate circles or ellipses to organize and show data.

2 To group together items that have something in common is to _____ the items.

3 The _____ is the middle number(s) in a set of data when the data are arranged in numerical order.

4 The number that appears most often in a set of data is called the _____.

5 A(n) _____ is a graph that is used to show how a set of data changes over a period of time.

6 _____ is information gathered for statistical purposes.

Label the diagram with the correct vocabulary term.

7 _____

8 _____

9 _____

10 _____

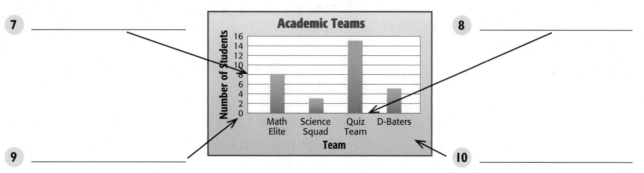

Lesson Review

7-1 Sort and Classify (pp. 274–280)

11 Create a Venn diagram to sort the numbers **4, 6, 7, 8, 9, 12,** and **14.** Classify them as multiples of 4 or odd numbers.

Multiples of 4: _____

Odd numbers: _____

Neither: _____

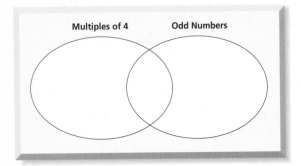

Example 1

Create a Venn diagram to sort the numbers 2, 3, 7, 8, 9, 11, and 12. Classify them as multiples of 3 or as prime numbers.

1. Multiples of 3: 9, 12

2. Prime numbers: 2, 7, 11

3. Both: 3

4. Neither: 8

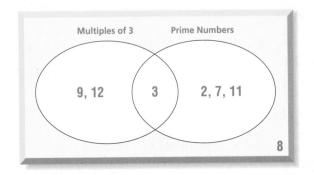

7-2 Mode, Median, and Range (pp. 281–288)

Find the mode and the range for each given set of data.

12 Jamari recorded the number of phone calls received each day.

12, 15, 17, 9, 15, 13, 11

The range is _____.

The mode is _____.

13 Mario recorded the number of hours he practiced piano each week.

5, 5, 7, 3, 7, 4, 7, 6, 3

The range is _____.

The mode is _____.

Example 2

Find the mode and the range for the given set of data.

Rafael recorded the number of siblings each of her teachers have.

2, 5, 2, 3, 4, 3, 1, 2, 0

1. Arrange the numbers in order from least to greatest.

0, 1, 2, 2, 2, 3, 3, 4, 5

2. The greatest number in the list is 5. The least number in the list is 0.

$$5 - 0 = 5$$

3. The range is 5.

4. The mode (the number listed most often) is 2.

7-3 Mean (pp. 289–295)

Find one missing number from a data set when the mean is given.

14 Mean: 6   Data set: 8, 3, 6, _____

15 Mean: 14   Data set: 13, 18, 10, _____

16 Mean: 27   Data set: 28, 28, 26, _____

17 Mean: 4.1
Data set: 1.9, 2.8, 6.9, _____

Example 3

The mean of five numbers is 6. Four of the numbers are 3, 7, 8 and 5. Find the missing number.

1. Find the total value of 5 numbers with a mean of 6.

  $5 \cdot 6 = 30$

2. Find the sum of the given numbers.

  $3 + 7 + 8 + 5 = 23$

3. Subtract the sum of the four numbers from the total sum.

  $30 - 23 = 7$

4. The missing number is 7.

7-4 Interpret Bar Graphs (pp. 297–304)

Use the double-bar graph "Favorite Spirit Day Apparel" to compare data.

18 Which apparel is preferred the least by boys?

_____

19 Which apparel is preferred the most by boys?

_____

20 How many more girls prefer face painting over T-shirts?

_____

21 How many more boys than girls prefer T-shirts?

_____

Example 4

Use the double-bar graph "Favorite Spirit Day Apparel" to compare data.

How many more girls chose face painting than hair accessories as their favorite Spirit Day apparel?

1. How many girls chose face painting? **11**

2. How many girls chose hair accessories? **5**

3. Subtract. **11 − 5 = 6**

4. Six more girls chose face painting.

Example 5

Use the data in the table to create a bar graph.

Pet Adopt, Inc. took a survey of 500 pet owners to see what type of pets they own.

Pet Adopt, Inc. Survey	
Type of Pet	Number Owned
Dog	300
Horse	25
Cat	100
Fish	50
Bird	25

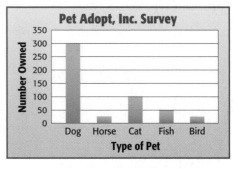

Use the data in the table to create a bar graph.

22 Starr City Library recorded the items that were checked out last week.

Starr City Library Circulation	
Item	Number Checked Out
Books	300
Audio Tapes	50
CDs	150
Books on Tape	100
DVDs	175

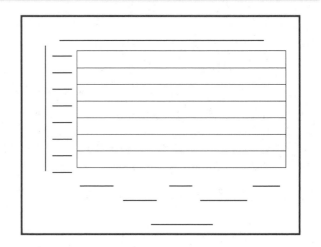

7-6 **Interpret Line Graphs** (pp. 315–322)

Use the double-line graph "Graphic T-Shirt Sales" to compare data.

23 Compare rock group T-shirt sales in June and April.

_____

24 Describe rock T-shirt sales in July.

_____

25 Compare sales of sports team T-shirts to rock group T-shirts in April.

_____

Example 6

Use the double-line graph "Graphic T-Shirt Sales" to compare data.

1. Rock group T-shirt sales in May were $15,000. The sport team T-shirt sales in May were $17,000.

2. Sales were $2,000 higher for sport team T-shirts.

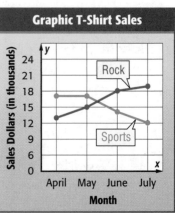

7-7 Create Line Graphs (pp. 323–329)

Example 7

Use the data in the table to create a line graph.

Mr. Murphy recorded the gallons of paint his company used over a four-month period.

Murphy Home Improvement	
Month	Gallons of Paint
November	95
December	52
January	37
February	25

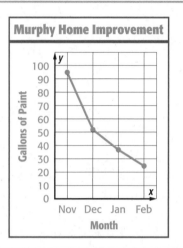

Use the data in the table to create a line graph.

26 Ms. Morales recorded the amount of money earned at Super-Fixer Auto Shop over a four-month period.

Super-Fixer Auto Shop	
Month	Money Earned
April	$72,000
May	$98,000
June	$81,000
July	$89,000

Use the data in the table to create a double-line graph.

27 Mr. Brooks recorded the sales trends of books and audio/visual products that his bookstore sold over a four-month period.

Brooks' Books Sales		
Month	Sales (in thousands)	
	Books	Audio/Visual
March	$12	$27
April	$14	$32
May	$16	$30
June	$19	$35

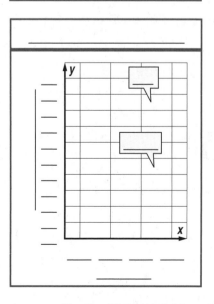

List the numbers in each category.

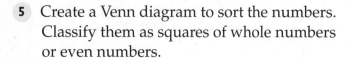

1   factors of 6 _____

2   odd numbers _____

3   both _____

4   neither _____

5   Create a Venn diagram to sort the numbers. Classify them as squares of whole numbers or even numbers.

3, 4, 9, 12, 25, 34, 36, 44, 49, 51

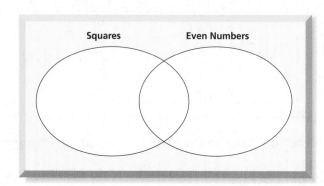

Squares   Even Numbers

Find the mode for each given set of data.

6   23, 17, 31, 39, 31 _____

7   97, 84, 57, 66, 97, 75, 97, 81, 68 _____

Find the median and the range for each given set of data.

8   Bianca asked seven students how many bedrooms are in their home.

3, 5, 2, 3, 2, 2, 4

The median is _____.   The range is _____.

9   George asked seven adults how many times they have moved.

5, 6, 1, 7, 6, 4, 12

The median is _____.   The range is _____.

Find the mean of the data set. Convert the remainder into a fraction or a decimal.

10   11, 9, 5, 7, 8, 5 _____

11   8, 11, 13, 9, 12, 10 _____

Find one missing number from a data set when the mean is given.

12 Mean: 9   Data set: 5, 12, 8, _____

13 Mean: 12   Data set: 11.4, 9.7, 10.8, _____

Use the bar graph "Betsy's Reading Log" to compare data.

14 What interval is used for the scale?

_____

15 What does the height of each bar represent?

_____

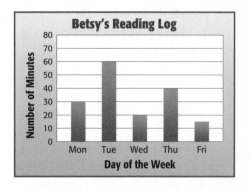

Use the data in the table to create a double-line graph.

16 DETERGENT The table shows sales of two different laundry detergent brands at Save-Mart over a four-month period.

Save-Mart Sales		
Month	Laundry Detergent	
	Value	Brite
March	$75	$175
April	$125	$200
May	$125	$225
June	$175	$250

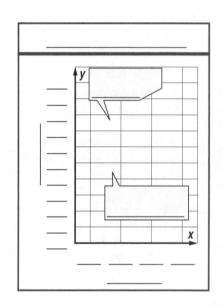

Correct the mistakes.

17 VOTING Khalil used the table below to make a double-bar graph. What was his mistake?

Student Council President Results		
Candidate	Number of Votes	
	Grade 7	Grade 8
Esmerelda	35	37
Minho	39	46
Demitri	46	23
Karena	29	38

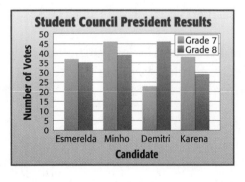

_____

_____

STOP

Choose the best answer and fill in the corresponding circle on the sheet at right.

1 Francisco recorded the ages of 7 of his teachers. What is the mode of this data set?

32, 53, 23, 24, 31, 44, 23

A 32.9 C 30

B 31 D 23

2 Miles recorded the number of students who play brass instruments in the school band. The mean is 6. What is the missing number of trumpet players in the table below?

Marcell Middle School Brass Section	
Instrument	Number of Players
Trumpet	
Trombone	6
Mellophone	3
Sousaphone	5

A 6 players C 12 players

B 10 players D 24 players

3 Kelly recorded the number of pounds of bananas sold each hour. What is the median number of pounds?

54, 52, 54, 53, 56, 57, 59, 51, 58

A 8 C 54.5

B 54 D 59

4 Use the bar graph, "Favorite Drinks" to compare data. How many more boys than girls chose chocolate milk as their favorite drink?

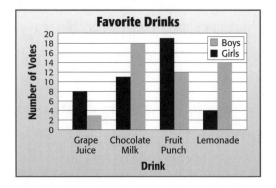

A 5 boys C 9 boys

B 7 boys D 10 boys

5 The Census Bureau recorded the number of people that live in each house on Pine Street. What is the range of this data set?

9, 4, 6, 3, 6, 2, 4, 3, 3, 5, 7

A 3 C 5

B 4 D 7

6 Mrs. Marcucci recorded Alexa's last seven quiz grades. What is the mean of this data set?

65, 77, 87, 91, 96, 77, 74

A 26 C 77

B 65 D 81

7 Use the line graph, "Centertown Shipping" to compare data. How many more staffing hours were used at 1 P.M. than at 10 A.M.?

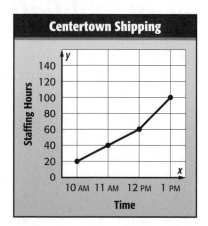

A 20 hours

B 80 hours

C 100 hours

D 120 hours

---

8 Use the double-line graph, "Bayside Boat Sales" to compare data. How much more was sold in April 2009 than in April 2008?

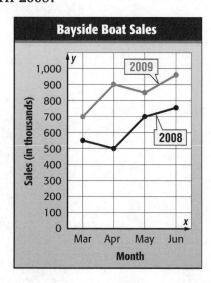

A $400,000

B $500,000

C $900,000

D $150,000

9 Beatriz is sorting a set of numbers into categories. In which category does the number 36 belong?

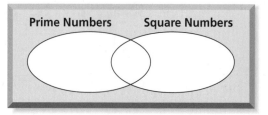

A prime numbers   **C** both

B square numbers   **D** neither

ANSWER SHEET

Directions: Fill in the circle of each correct answer.

1 Ⓐ Ⓑ Ⓒ Ⓓ

2 Ⓐ Ⓑ Ⓒ Ⓓ

3 Ⓐ Ⓑ Ⓒ Ⓓ

4 Ⓐ Ⓑ Ⓒ Ⓓ

5 Ⓐ Ⓑ Ⓒ Ⓓ

6 Ⓐ Ⓑ Ⓒ Ⓓ

7 Ⓐ Ⓑ Ⓒ Ⓓ

8 Ⓐ Ⓑ Ⓒ Ⓓ

9 Ⓐ Ⓑ Ⓒ Ⓓ

Success Strategy

If two answers seem correct, compare them for differences. Reread the problem to find the best answer between the two.

Chapter 8

Percents and Circle Graphs

Percents can be used to compare values.

For example, 70% of the Earth's surface is covered with water. Surveys often show results in percents, such as 90% of dentists prefer a certain toothbrush. School systems also use percents to show the number of students who have graduated.

STEP **1** Quiz

Math Online > Are you ready for Chapter 8? Take the Online Readiness Quiz at *glencoe.com* to find out.

STEP **2** Preview

Get ready for Chapter 8. Review these skills and compare them with what you will learn in this chapter.

What You Know	What You Will Learn

What You Know

You understand the meaning of fractions.

Examples:

$\frac{1}{2}$

$\frac{5}{10}$

TRY IT!

Identify the fractions below.

2

_____   _____

You know how to use equal size pieces to show a fraction.

not a fraction   fraction: $\frac{3}{4}$

TRY IT!

Use the circle to show each fraction.

3   **4**

_____   _____

What You Will Learn

Lesson 8-1

A **percent** is a ratio that compares a number to 100. You can use a fraction with a denominator of 100 to find a percent.

$$\frac{45}{100} = 45\%$$

$$\frac{27}{100} = 27\%$$

Percents can also be written as ratios and decimals.

45% means 45 out of 100 or 0.45

Lesson 8-2

Each **sector** in a **circle graph** does not need to be the same size. However, the sum of the percents shown by each sector equals 100%.

Favorite Colors

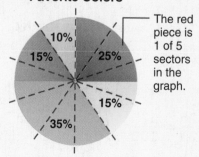

The red piece is 1 of 5 sectors in the graph.

10% 25% 15% 15% 35%

Percents

KEY Concept

A **percent** is a ratio that compares a number to 100. A percent is written using the percent symbol (%).

The word *percent* means "hundredths" or "out of 100." Percents can be written as fractions or decimals because they show the relationship between one whole (100%) and a part of a whole.

Describe the shaded area in different ways.

Ratio: 30 out of 100

Fraction: $\dfrac{30}{100}$

Percent: 30%

Decimal: 0.30

The fraction above could also be simplified to find an **equivalent fraction**.

$$30\% = \frac{30 \div 10}{100 \div 10} = \frac{3}{10}$$

Copyright © Glencoe/McGraw-Hill, a division of The McGraw-Hill Companies, Inc.

VOCABULARY

equivalent fractions
fractions that name the same number

percent
a ratio that compares a number to 100

ratio
a comparison of two numbers by division

You can use models to show the relationships between fractions and percents.

Example 1

Identify the percent that is modeled.

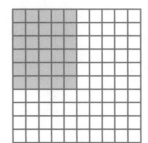

1. Find the fraction that is shaded.

$1\dfrac{1}{4}$ or $\dfrac{5}{4}$

2. Find an equivalent fraction with a denominator of 100.

$$\frac{5 \cdot 25}{4 \cdot 25} = \frac{125}{100}$$

3. Write the percent.

125%

YOUR TURN!

Identify the percent that is modeled.

1. Find the fraction that is shaded.

   _____

2. Find an equivalent fraction with a denominator of 100.

   $$\frac{\boxed{} \cdot \boxed{}}{\boxed{} \cdot \boxed{}} = \frac{\boxed{}}{100}$$

3. Write the percent.

   _____

Example 2

Find the missing value. Solve.

What is 25% of 8?

1. Write the percent as a decimal.

   25% = 0.25

2. Multiply the decimal and the whole number.

   0.25 · 8 = 2

YOUR TURN!

Find the missing value. Solve.

What is 40% of 60?

1. Write the percent as a decimal.

   40% = _____

2. Multiply the decimal and the whole number.

   _____ · _____ = _____

Who is Correct?

Write $\frac{12}{50}$ as a percent.

Dehlila
$$\frac{12 \cdot 2}{50 \cdot 2} = \frac{24}{100} = 24\%$$

Jordan
$$\frac{12 \div 2}{50 \div 2} = \frac{6}{25} = 6\%$$

Perry
$$\frac{12 + 50}{50 + 50} = \frac{62}{100} = 62\%$$

Circle correct answer(s). Cross out incorrect answer(s).

GO ON

Copyright © Glencoe/McGraw-Hill, a division of The McGraw-Hill Companies, Inc.

Lesson 8-1 Percents   **343**

▶ Guided Practice

Identify each percent that is modeled.

1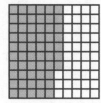

_____

2

_____

Step (by) Step Practice

3 Identify the percent of green buttons in the total buttons.

Step 1 Write the percent of green buttons to total buttons.

$$\frac{\text{green buttons}}{\text{total buttons}} = \text{_____}$$

Step 2 Find an equivalent fraction with a denominator of 100.

$$\frac{\square}{\square} \cdot \frac{\square}{\square} = \frac{\square}{100}$$

Step 3 Write the percent as a decimal. _____

Step 4 Write the percent. _____

Identify each percent that is modeled.

4

fraction: _____

fraction with a denominator of 100:

$$\frac{\square}{\square} \cdot \frac{\square}{\square} = \frac{\square}{100}$$

decimal: _____

percent: _____

5

fraction: _____

fraction with a denominator of 100:

$$\frac{\square}{\square} \cdot \frac{\square}{\square} = \frac{\square}{100}$$

decimal: _____

percent: _____

Find the missing value.

6 What is 30% of 20?

30% = _____

_____ • _____ = _____

7 What is 95% of 420?

95% = _____

_____ • _____ = _____

8 What is 110% of 46?

110% = _____

_____ • _____ = _____

9 What is 86% of 86?

86% = _____

_____ • _____ = _____

Step by Step Problem-Solving Practice

Solve.

10 **ELECTIONS** There are 300 students who voted for class president. Marcell won 34% of the vote, Antoinette won 29% of the vote, and Miranda won 37% of the vote. How many votes did each candidate receive?

Problem-Solving Strategies
☑ Use a table.
☐ Draw a diagram.
☐ Use logical reasoning.
☐ Solve a simpler problem.
☐ Work backward.

Understand Read the problem. Write what you know.

Marcell won _____ of the vote.
Antoinette won _____ of the vote.
Miranda won _____ of the vote.

Plan Pick a strategy. One strategy is to make a table.

Solve Complete the table to find the number of votes for each candidate.

To find the number of votes, write the percent as a _____.

Then, multiply the _____ times the _____.

Name	Percent	Calculation	Number of Votes
Marcell	%	• =	
Antoinette	%	• =	
Miranda	%	• =	

Check Add the number of votes for each candidate. The sum should equal 300 votes.

GO ON

Solve.

11 **SURVEYS**   The Sweet Tooth Ice Cream Company took a survey of 250 customers' favorite flavors. Forty-eight percent chose vanilla, twenty-four percent chose chocolate, and twenty-eight percent chose strawberry as their favorite. How many people chose each flavor?

Check off each step.

_____ **Understand: I underlined key words.**

_____ **Plan: To solve this problem, I will** _____.

_____ **Solve: The answer is** _____.

_____ **Check: To check my answer, I will** _____

_____.

12 **FINANCES**   Mrs. Arnold earns $2,075 dollars a month. If 40% of her salary is spent on rent, how much does she pay in rent each month?

_____

13 **Reflect**   Roald read 15 out of 25 books on his bookshelf. Explain how to find the percentage of books on the bookshelf that he has read.

_____

 Skills, Concepts, and Problem Solving

Identify each percent that is modeled.

14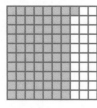

_____

15

_____

Identify each percent that is modeled.

16

fraction: _____

fraction with a denominator of 100:

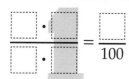
$$\frac{\square \cdot \square}{\square \cdot \square} = \frac{\square}{100}$$

decimal: _____

percent: _____

17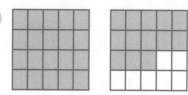

fraction: _____

fraction with a denominator of 100:

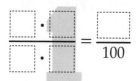
$$\frac{\square \cdot \square}{\square \cdot \square} = \frac{\square}{100}$$

decimal: _____

percent: _____

18

fraction: _____

fraction with a denominator of 100:

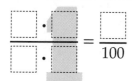
$$\frac{\square \cdot \square}{\square \cdot \square} = \frac{\square}{100}$$

decimal: _____

percent: _____

19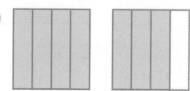

fraction: _____

fraction with a denominator of 100:

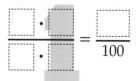
$$\frac{\square \cdot \square}{\square \cdot \square} = \frac{\square}{100}$$

decimal: _____

percent: _____

Find the missing value.

20 What is 5% of 80?

_____ • _____ = _____

21 What is 165% of 260?

_____ • _____ = _____

22 What is 40% of 79?

_____ • _____ = _____

23 What is 17% of 34?

_____ • _____ = _____

24 What is 81% of 260?

_____ • _____ = _____

25 What is 47% of 928?

_____ • _____ = _____

GO ON

Solve.

26 **RAINFALL** On average, about 38 inches of precipitation falls in Arkansas City, Kansas. About 26% of the precipitation occurs in May and June. About how much precipitation occurs within this time period? Round to the nearest whole number.

_____

27 **BUDGETS** Mr. Blackwell earns $1,275.00 each week. He budgets 3% of his earnings for his electric bill. How much money does Mr. Blackwell budget each week for his electric bill?

_____

28 **FINANCIAL LITERACY** Alexis earned $57.00 this month. She has decided to deposit 10% of her earnings into a savings account. How much money will she deposit into savings?

_____

29 **MUSEUMS** Twenty-seven percent of the paintings at the Archibald Museum of Art are watercolors. If the museum has a collection of 400 paintings, how many watercolors are in the museum?

_____

Vocabulary Check **Write the vocabulary word that completes each sentence.**

30 A(n) _____ is a ratio that compares a number to 100.

31 The numbers $\frac{2}{5}$ and $\frac{40}{100}$ are _____.

32 **Writing in Math** Explain how to write the mixed number $1\frac{47}{50}$ as a percent.

_____

_____

_____

Lesson 8-2 Percents and Angle Measures

KEY Concept

The relationship between percents, decimals, and fractions can be shown with a circle. Every circle is made of a total of 360°. The circle can be divided into pie-shaped sections called sectors.

The circle below shows 20 sectors of equal size. Each sector represents 5% of the circle.

Percent: 5%

Decimal: 0.05

Fraction: $\frac{5}{100}$

Simplest Form: $\frac{1}{20}$

To find the degree measure of each sector in the circle on the right, multiply the total degrees of the circle (360°) by the fraction in simplest form.

$$\frac{360°}{1} \cdot \frac{1}{20} = \frac{360°}{20} = 18°$$

Each 5% sector has a measure of 18°.

Another way to find the degreee measure of a 5% sector is to multiply 360° by the decimal 0.05.

$$360° \cdot 0.05 = 18°C$$

VOCABULARY

degree
the most common unit of measure for an angle

denominator
the bottom number in a fraction; it represents the number of parts in the whole

percent
the ratio that compares a number to 100

sector
pie-shaped sections in a circle

simplest form
the form of a fraction when the GCF of the numerator and the denominator is 1

Common percents, such as 5%, 10%, 25%, and 50%, can be used in combinations to make other useful percents and circle graphs.

Example 1

Find the degrees needed to show a 10% sector in a circle graph.

1. Write the percent as a fraction in simplest form.

$$10\% = \frac{10}{100} \div \frac{10}{10} = \frac{1}{10}$$

$\frac{1}{10}$ of the circle is shaded.

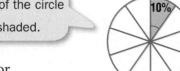

2. Use the fraction to find the degree measure of a 10% sector.

$$\frac{360°}{1} \cdot \frac{1}{10} = \frac{360°}{10} = 36°$$

3. Check by multiplying 360° by the percent in decimal form.

$$360° \cdot 10\% = 360° \cdot 0.10 = 36°$$

GO ON

YOUR TURN!

Find the degrees needed to show a 25% sector in a circle graph.

1. Write the percent as a fraction in simplest form.
   (Hint: What fraction of the circle is shaded?)

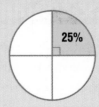

25% = □/100 ÷ □/□ = □/□

2. Use the fraction to find the degree measure of a 25% sector.

□/1 · □/□ = □/□ =

3. Check by multiplying 360° by the percent in decimal form.

360° · 25% = 360° · _____ = _____

Example 2

Use combinations to find the degrees needed to show a 35% sector in a circle graph.

1. Write 35% as a combination of percents.

   5% + 10% + 10% + 10% = 35%

2. Find each percent in sector degrees.

   $5\% = \dfrac{360}{1} \cdot \dfrac{5}{100} = \dfrac{1800}{100} = 18°$

   $10\% = \dfrac{360}{1} \cdot \dfrac{10}{100} = \dfrac{3600}{100} = 36°$

3. Find the sum of the sector degrees.

   5% + 10% + 10% + 10% = 35%

   18° + 36° + 36° + 36° = 126°

YOUR TURN!

Use combinations to find the degrees needed to show a 60% sector in a circle graph.

1. Write 60% as a combination of percents.

   _____ + _____ + _____ = _____

2. Find each percent in sector degrees.

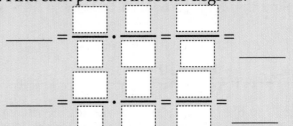

   _____ = □/□ · □/□ = □/□ = _____

   _____ = □/□ · □/□ = □/□ = _____

3. Show each percent in sector degrees.

   _____ + _____ + _____ = _____

   _____ + _____ + _____ = _____

Who is Correct?

Find the degrees needed to show a 50% sector in a circle graph.

Andrés
$$50\% = \frac{1}{2}$$
$$\frac{360°}{1} \cdot \frac{1}{2} = \frac{360°}{2} = 180°$$

Umeko
$$50\% = \frac{1}{50}$$
$$\frac{360°}{1} \cdot \frac{1}{50} = \frac{360°}{50} = 7.2°$$

Alexander
$$50\% = \frac{1}{2}$$
$$\frac{360°}{1} \cdot \frac{1}{2} = \frac{360°}{50} = 358°$$

Circle the correct answer(s). Cross out incorrect answer(s).

▶ Guided Practice

Name the fraction and the percent of the circle that is shaded.

1

fraction: _____

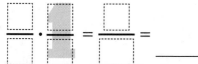

2

fraction: _____

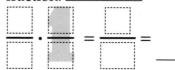

Step by Step Practice

3 Find the degrees needed to show a 20% sector in a circle graph.

Step 1 Write the percent as a fraction in simplest form.

$$20\% = \frac{\boxed{}}{100} \div \frac{\boxed{}}{\boxed{}} = \frac{\boxed{}}{\boxed{}}$$

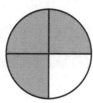
20%

Step 2 Use the simplest form to identify the number of sectors created.

The denominator is _____, so there are _____ sectors created.

Step 3 Find the degree measure of a 20% sector.

$$\frac{\boxed{}}{1} \cdot \frac{\boxed{}}{\boxed{}} = \frac{\boxed{}}{\boxed{}} = \underline{}$$

GO ON ▶

Copyright © Glencoe/McGraw-Hill, a division of The McGraw-Hill Companies, Inc.

Find the degrees needed to show each sector in a circle graph.

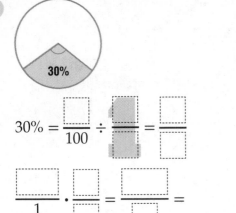

4

$$30\% = \frac{\boxed{}}{100} \div \frac{\boxed{}}{\boxed{}} = \frac{\boxed{}}{\boxed{}}$$

$$\frac{\boxed{}}{1} \cdot \frac{\boxed{}}{\boxed{}} = \frac{\boxed{}}{\boxed{}} = \underline{}$$

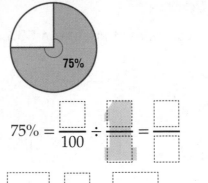

5

75%

$$75\% = \frac{\boxed{}}{100} \div \frac{\boxed{}}{\boxed{}} = \frac{\boxed{}}{\boxed{}}$$

$$\frac{\boxed{}}{1} \cdot \frac{\boxed{}}{\boxed{}} = \frac{\boxed{}}{\boxed{}} = \underline{}$$

Use the answers to the previous exercises to complete the chart below.

6

Percent	Number of Sectors	Degree Measure of Each Sector
5		
10		
20		
25		
50		
100		

Use combinations from the values in the chart to find the degrees needed to show each sector in a circle graph.

7 15%

_____ + _____ = _____

_____ + _____ = _____

8 85%

_____ + _____ + _____ = _____

_____ + _____ + _____ = _____

9 45%

_____ + _____ = _____

_____ + _____ = _____

10 65%

_____ + _____ + _____ = _____

_____ + _____ + _____ = _____

Step by Step Problem-Solving Practice

Problem-Solving Strategies
☐ Use a table.
☐ Draw a diagram.
☐ Use logical reasoning.
☑ Use a formula.
☐ Work backward.

11 ELECTIONS Nita is writing an article on the recent election of the student body president. She includes a circle graph that shows how each class contributed to the voting. Of the votes cast, 30% were by 6th graders, 45% by 7th graders, and the rest by 8th graders. Find the degree measure of each sector in the circle graph.

Understand   Read the problem. Write what you know.

The circle must show _____ for 6th graders,

_____ for 7th graders, and _____ for 8th graders.

Plan   Pick a strategy. One strategy is to use a formula. Use the decimal form of each percent to find the number of degrees in each sector.

Solve   6th graders: 360° · _____ = 360° · _____ = _____

7th graders: 360° · _____ = 360° · _____ = _____

8th graders: 360° · _____ = 360° · _____ = _____

Check   The sum of the degree measures should equal 360°.

_____ + _____ + _____ = _____

12 SURVEYS The eighth grade students voted on their choices for a class trip. The circle graph shows their choices. What is the degree measure of each sector in the circle graph?

Check off each step.

_____ Understand: I underlined key words.

_____ Plan: To solve the problem, I will _____.

_____ Solve: The answer is _____

_____.

_____ Check: I checked my answer by _____.

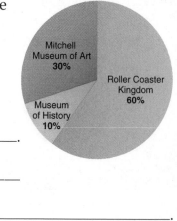

Class Trip

Mitchell Museum of Art **30%**

Roller Coaster Kingdom **60%**

Museum of History **10%**

GO ON

13 **LOGOS** Hi-Five Furniture wants to design a logo using a circle with 5 sectors of equal size. What percentage of the circle will each sector represent? What is the degree measure of each sector?

_____

_____

14 **Reflect** A 5% sector of a circle has a degree measure of 18°. How can you use this information to find the degree measure of a 25% sector?

_____

_____

_____

▶ Skills, Concepts, and Problem Solving

Name the fraction and the percent of the circle that is shaded.

15

fraction: _____

$$\frac{\square}{\square} \cdot \frac{\square}{\square} = \frac{\square}{\square} = \underline{\quad}$$

16

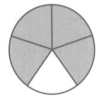

fraction: _____

$$\frac{\square}{\square} \cdot \frac{\square}{\square} = \frac{\square}{\square} = \underline{\quad}$$

Find the degrees needed to show each sector in a circle graph.

17

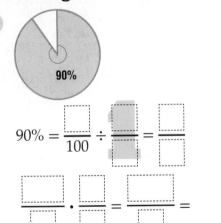

90%

$$90\% = \frac{\square}{100} \div \frac{\square}{\square} = \frac{\square}{\square}$$

$$\frac{\square}{\square} \cdot \frac{\square}{\square} = \frac{\square}{\square} = \underline{\quad}$$

18

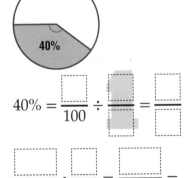

40%

$$40\% = \frac{\square}{100} \div \frac{\square}{\square} = \frac{\square}{\square}$$

$$\frac{\square}{\square} \cdot \frac{\square}{\square} = \frac{\square}{\square} = \underline{\quad}$$

Use the chart from Exercise 6 to find the degrees needed to show each sector.

19 35%

_____ + _____ = _____

_____ + _____ = _____

20 80%

_____ + _____ + _____ = _____

_____ + _____ + _____ = _____

Solve.

21 SUMMER SURVEY Mary Beth surveyed 250 students about their favorite summer vacation activity. Complete the table and find the degree measure in a circle graph needed to display each survey answer.

Activity	Responses	Degree Measure
Family Vacation	20%	
Outdoor Activities	45%	
Sleeping In	35%	

Vocabulary Check **Write the vocabulary word that completes each sentence.**

22 A(n) _____ is a pie-shaped section in a circle.

23 The _____ of $\frac{25}{100}$ is $\frac{1}{4}$.

24 Writing in Math Explain how you could use subtraction and commonly known percents to find the degree value of a 70% sector.

_____

_____

_____

 Spiral Review

Solve. (Lesson 8-1, p. 342)

25 CHARITY Maxwell earned $878.00 this month. He has decided to give 10% of his earnings to a local charity. How much money will he donate?

_____

Progress Check 1 (Lessons 8-1 and 8-2)

Identify the percent that is modeled.

1
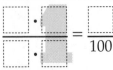

fraction: _____

fraction with a denominator of 100:

$$\frac{\boxed{} \cdot \boxed{}}{\boxed{} \cdot \boxed{}} = \frac{\boxed{}}{100}$$

decimal: _____

percent: _____

2
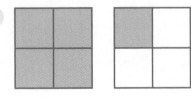

fraction: _____

fraction with a denominator of 100:

$$\frac{\boxed{} \cdot \boxed{}}{\boxed{} \cdot \boxed{}} = \frac{\boxed{}}{100}$$

decimal: _____

percent: _____

Find the missing value. Round your answer to the nearest hundredth.

3 What is 7% of 210?

_____ · _____ = _____

4 What is 23% of 241?

_____ · _____ = _____

Find the degrees needed to show each sector in a circle graph.

5
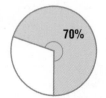
70%

$$70\% = \frac{\boxed{}}{100} \div \frac{\boxed{}}{\boxed{}} = \frac{\boxed{}}{\boxed{}}$$

$$\frac{\boxed{}}{1} \cdot \frac{\boxed{}}{\boxed{}} = \frac{\boxed{}}{\boxed{}} = \underline{}$$

6
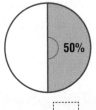
50%

$$50\% = \frac{\boxed{}}{100} \div \frac{\boxed{}}{\boxed{}} = \frac{\boxed{}}{\boxed{}}$$

$$\frac{\boxed{}}{1} \cdot \frac{\boxed{}}{\boxed{}} = \frac{\boxed{}}{\boxed{}} = \underline{}$$

Solve.

7 **FINANCIAL LITERACY** Mr. Matthews had $3,000 in his savings account. If he withdrew 25%, how much did he withdraw?

_____

Interpret Circle Graphs

KEY Concept

One way to represent data is in a **circle graph**. A circle graph has pie-shaped sections called **sectors**. Sectors can be used to compare the parts of a whole.

The circle graph shows the favorite sports of 200 students at Lincoln Middle School.

The whole circle represents the total of all the data, or 100% of the data. The sectors of the graph equal 100%.

Favorite Sports

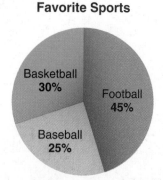

Basketball 30%

Football 45%

Baseball 25%

$$45\% + 25\% + 30\% = 100\%$$

The table shows the relationship between the data that was collected and the percent in the chart.

Sport	Percent	Fractions	Number of Students
Football	45%	$\frac{45}{100} = \frac{90}{200}$	90
Baseball	25%	$\frac{25}{100} = \frac{50}{200}$	50
Basketball	30%	$\frac{30}{100} = \frac{60}{200}$	60

The percent equation can also be used to find the number of students who chose each sport.

$$\text{percent} \cdot \text{whole} = \text{part}$$

$$0.45 \cdot 200 = 90$$

The equation above shows that 90 students, out of 200, represents 45% of those surveyed.

VOCABULARY

circle graph
a graph used to compare parts of a whole; the circle represents the whole and is separated into parts of a whole

data
information, often numerical, which is gathered for statistical purposes

degree
the most common unit of measure for angles

percent
a ratio that compares a number to 100

sector
pie-shaped sections in a circle graph

To interpret the information in the circle graph above, it is important to know that 200 students were surveyed. This information is needed to find the number of responses to each choice.

GO ON

Example 1

Last month 100 animals were adopted from the Central Animal Shelter. Which two animals together were adopted as often as cats?

Pets Adopted

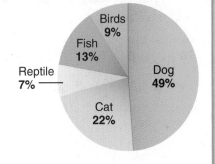

1. What percent of the animals adopted were cats? **22%**

2. What two categories equal 22% when added together?

   Reptile + Fish = 7% + 13% = 20%

   Reptile + Bird = 7% + 9% = 16%

   Fish + Bird = 13% + 9% = 22%

3. Fish and birds were adopted as often as cats.

YOUR TURN!

Big Sale Books took an inventory of the last 100 books that were sold. Which type of books was sold half as often as mysteries?

Books Sold

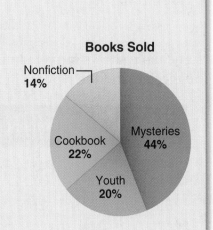

1. What percent of the books sold were mysteries? _____

2. What percent is equal to half of _____?

   _____ $\cdot \dfrac{1}{2} =$ _____

3. _____ were sold _____ of the time, or half as often as mysteries.

Example 2

Coconut Grove Consignment Shop sold 200 items of clothing last month. How many items were blue jeans?

Clothing Sold

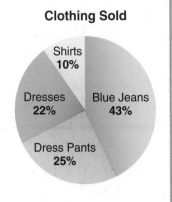

1. What percent of the items were blue jeans? **43%**

2. There were 200 total items sold. *This is the "whole."*

3. Use the percent equation to solve.

   percent • whole = part of clothing

   43% • 200 items = part of clothing   *Remember: 43% is the same as 0.43.*

   0.43 • 200 items = 86 blue jeans

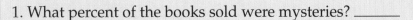

YOUR TURN!

Julieta spent $50 yesterday. How many dollars did Julieta spend on gasoline?

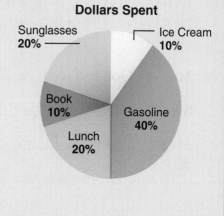

Dollars Spent

1. What percent of the money did she spend on gasoline? _____

2. Julieta spent a total of _____ dollars.

3. Use the percent equation to solve.

percent · whole = part

_____ · $_____ = part of dollars spent

_____ · $_____ = $_____

Who is Correct?

Refer to the circle graph above titled "Dollars Spent." How many dollars did Julieta spend on a book? Show your work.

Santos
percent · whole = part
0.10 · $100 = $10

Lakeisha
percent · whole = part
0.30 · $50 = $15

Carmen
percent · whole = part
0.10 · $50 = $5

 **Guided Practice**

The circle graph below shows the number of each type of home of 100 middle school teachers. Use the circle graph to answer the questions below.

1 What is the title of the graph?

_____

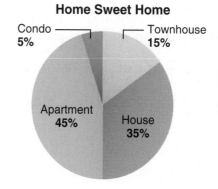

Home Sweet Home

2 What type of home do the greatest number of teachers live in?

_____

3 What percent of teachers do not live in a townhouse?

_____

 GO ON

4   At the county fair, 200 votes were cast for three different peach pie makers. The person with the most votes won the blue ribbon. How many votes did Oda receive?

Blue Ribbon Results

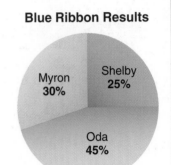

Step 1   Oda received _____ of the vote.

Step 2   There were _____ total votes cast.

Step 3   Use the percent equation to solve.

percent · whole = part

_____ · _____ = _____

Step 4   Use division to check your work.

_____ = _____

Big Sale Movies took an inventory of the last 100 movies that were sold. Use the circle graph to answer the questions below.

Movies Sold

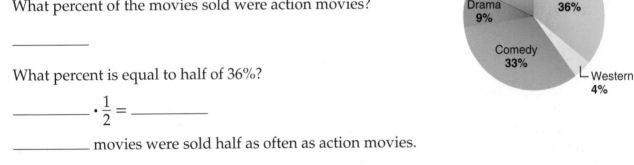

5   What type of movie was sold half as often as action films?

What percent of the movies sold were action movies?

_____

What percent is equal to half of 36%?

_____ · $\frac{1}{2}$ = _____

_____ movies were sold half as often as action movies.

The Fighting Falcons scored 100 points in last week's basketball game. Use the circle graph to answer the questions below.

Game Points

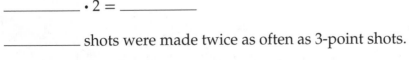

6   What type of shot was made twice as often as 3-point shots?

What percent of the points scored were from 3-point shots?

_____

What percent is two times 24%?

_____ · 2 = _____

_____ shots were made twice as often as 3-point shots.

Solve.

Problem-Solving Strategies
- ☐ Draw a diagram.
- ☑ Make a table.
- ☐ Work backward.
- ☐ Solve a simpler problem.
- ☐ Look for a pattern.

7 DANCES   Aisha is helping to plan the next school dance. In order to decide on a theme, she asks students which theme they like best: "70s Disco," "Barn Dance," or "Rock the Night."

A total of 300 students voted. How many students voted for the Barn Dance theme?

Understand   Read the problem and review the circle graph. Write what you know.

Favorite Dance Theme

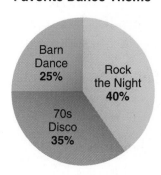

Barn Dance 25%
Rock the Night 40%
70s Disco 35%

There were _____ votes.

"Rock the Night" is the favorite of _____ of students.

"70s Disco" is the favorite of _____ of students.

"Barn Dance" is the favorite of _____ of students.

Plan   Pick a strategy. One strategy is to make a table.

Solve   Fill in the table.

percent of votes • total votes = votes for each theme

Theme	Rock the Night	70s Disco	Barn Dance
Percent			
Votes			

Check   Add the votes for each theme. Does the sum equal total votes for the dance?

GO ON

8 **PARTIES**   Colin's mom plans children's birthday parties. Last year she planned 400 parties. Use the circle graph to find the number of Petting Zoo parties she planned.

Check off each step.

_____ **Understand: I underlined key words.**

_____ **Plan: To solve this problem, I will** _____ .

_____ **Solve: The answer is** _____ .

_____ **Check: I checked my answer by** _____ .

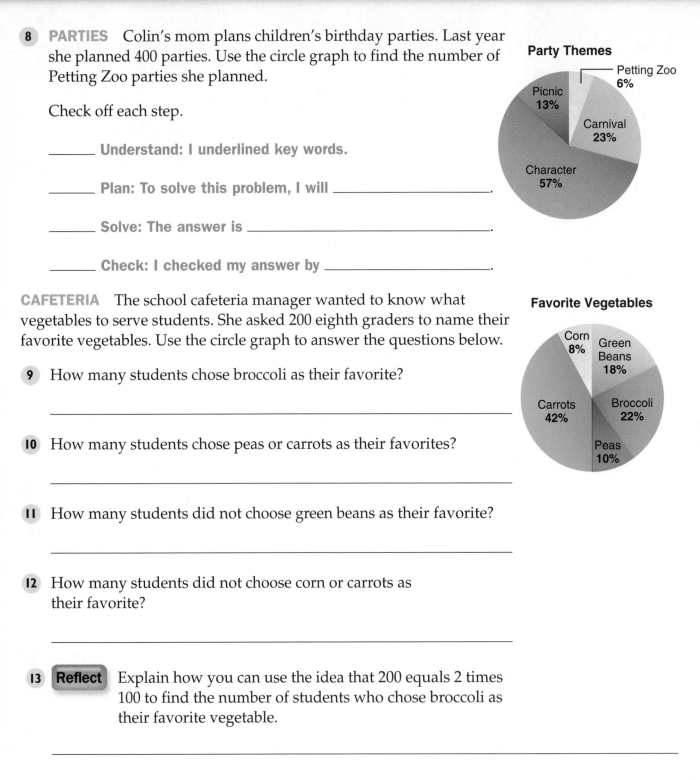

Party Themes

Petting Zoo 6%
Picnic 13%
Carnival 23%
Character 57%

CAFETERIA   The school cafeteria manager wanted to know what vegetables to serve students. She asked 200 eighth graders to name their favorite vegetables. Use the circle graph to answer the questions below.

Favorite Vegetables

Corn 8%
Green Beans 18%
Carrots 42%
Broccoli 22%
Peas 10%

9   How many students chose broccoli as their favorite?

_____

10   How many students chose peas or carrots as their favorites?

_____

11   How many students did not choose green beans as their favorite?

_____

12   How many students did not choose corn or carrots as their favorite?

_____

13   **Reflect**   Explain how you can use the idea that 200 equals 2 times 100 to find the number of students who chose broccoli as their favorite vegetable.

_____

_____

_____

Skills, Concepts, and Problem Solving

The circle graph at the right shows the percentage of tickets given to each group for a high school football championship. Use the circle graph to answer the questions below.

Championship Tickets

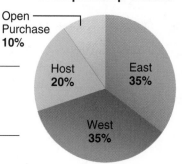

14 What is the title of the graph?

_____

15 What type of ticket will be sold the least?

_____

16 What percent of tickets will not go to the East or West division teams?

_____

Mr. Leapley used a circle graph to compare the number of students that played each instrument in the band. There are 200 students in the band. Use the circle graph to answer the questions below.

Instruments Played

17 What type of instrument is played by four times as many band members as the trumpet?

What percent of the band members play trumpet?

_____

What percent is equal to four times 9%?

_____ • 4 = _____

The _____ is played by four times as many band members.

18 Which two instruments are played by as many band members as the clarinet?

What two percents equal _____?

_____ + _____ = _____

The _____ and the _____ are played by as many members as the clarinet.

The clarinet is played by _____ of the band members.

GO ON

FIELD DAY MacMurray Middle School is planning the field day events. Fifty eighth graders were asked to name their favorite event. Use the circle graph to answer the questions below.

Favorite Field Day Event

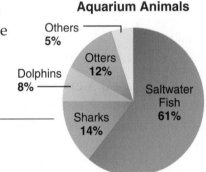

100-Meter Race 14%
Field Course Challenge 32%
400-Meter Relay 36%
Long Jump 18%

19 How many students chose the long jump as their favorite event?

_____

20 How many students chose the 100-Meter Race or the 400-Meter Relay as their favorites?

_____

21 How many students did not choose the Field Course Challenge as their favorite?

_____

AQUARIUMS Centerburg Aquarium has 500 animals. The circle graph compares the number of each type of animal. Use the circle graph to answer the questions below.

Aquarium Animals

Others 5%
Otters 12%
Dolphins 8%
Saltwater Fish 61%
Sharks 14%

22 How many saltwater fish are at the aquarium?

_____

23 How many dolphins and sharks are at the aquarium?

_____

24 How many animals are not otters?

_____

▶ Spiral Review

Find the degrees needed to show each sector in a circle graph. (Lesson 8-2, p. 349)

25

80%

$80\% = \dfrac{\square}{100} \div \dfrac{\square}{\square} = \dfrac{\square}{\square}$

$\dfrac{\square}{\square} \cdot \dfrac{\square}{\square} = \dfrac{\square}{\square} = \underline{\hspace{1cm}}$

26

25%

$25\% = \dfrac{\square}{100} \div \dfrac{\square}{\square} = \dfrac{\square}{\square}$

$\dfrac{\square}{\square} \cdot \dfrac{\square}{\square} = \dfrac{\square}{\square} = \underline{\hspace{1cm}}$

STOP

Create Circle Graphs

KEY Concept

To create a circle graph, it can be helpful to use data from a table. The table shows the relationship between the data and the degree measures of the sectors in a circle graph.

Favorite Vacation Spots

Vacation Spots	Number of People	Percent Value	Degree Measure
Beach	15	15%	$0.15 \cdot 360 = 54°$
Camping	10	10%	$0.10 \cdot 360 = 36°$
Water Park	30	30%	$0.30 \cdot 360 = 108°$
Amusement Park	35	35%	$0.35 \cdot 360 = 126°$
Grand Canyon	10	10%	$0.10 \cdot 360 = 36°$
Total	**100**	**100%**	**360°**

To make the graph, start by dividing the circle into 10 equal pieces. Each mark represents 36°, or 10% of the circle.

Shade and label each sector working around the circle clockwise. To show 5%, cut a 10% section in half.

Favorite Vacation Spots

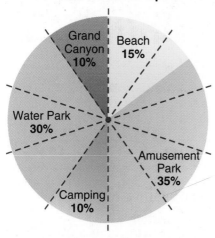

VOCABULARY

center
the given point from which all points on a circle are the same distance

circle graph
a graph used to compare parts of a whole; the circle represents the whole and is separated into parts of a whole

data
information, often numerical, which is gathered for statistical purposes

degree
the most common unit of measure for angles

percent
a ratio that compares a number to 100

sector
pie-shaped sections in a circle graph

You can build a ratio table that provides a quick reference for many common percents and their degree measures.

GO ON

Example 1

Complete the table to show the decimal value, the percent, and the degree measure for each type of lunch.

Favorite Lunch	Number of People	Decimal Value	Percent Value	Degree Measure
Spaghetti	375	375 ÷ 1,500 = 0.25	25%	90°
Enchiladas	225	225 ÷ 1,500 = 0.15	15%	54°
Sloppy Joes	525	525 ÷ 1,500 = 0.35	35%	126°
Taco Salad	225	225 ÷ 1,500 = 0.15	15%	54°
Veggie Burgers	150	150 ÷ 1,500 = 0.10	10%	36°
Total	1,500	1.00	100%	360°

1. To write the decimal, divide the number for each lunch choice by the total surveyed.

2. Find the equivalent percent value for each decimal.

3. Change all percent values to degree measures.
   For example, 25% of 360° = 90°.

YOUR TURN!

Complete the table to show the decimal value, the percent, and the degree measure for each of Javier's income sources.

Source of Income	Dollars Earned	Decimal Value	Percent Value	Degree Measure
Babysitting	$555			
Summer Job	$1,850			
After School Tutoring	$1,110			
Holiday Gifts	$185			
Total	3,700	1.00	100%	360°

1. To write the decimal, divide the number for each source of income by the total dollars earned.

2. Find the equivalent percent value for each decimal.

3. Change all percent values to degree measures.

Example 2

Use the table in Example 1 on page 366 to create a circle graph.

1. Write a title for the graph.

2. Draw sector lines using 10% marks as guides.

3. Label the sectors of the circle for each category.

4. Color each sector a different color. Write the percent value in each sector.

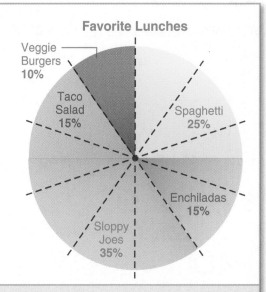

Favorite Lunches

Veggie Burgers 10%
Taco Salad 15%
Spaghetti 25%
Enchiladas 15%
Sloppy Joes 35%

YOUR TURN!

Use the table from Your Turn on page 366 to create a circle graph.

1. Write a title for the graph.

2. Draw sector lines using 10% marks as guides.

3. Label the sectors of the circle for each source of income.

4. Color each sector a different color. Write the percent value in each sector.

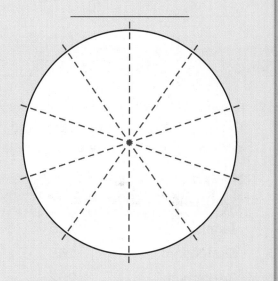

Who is Correct?

A survey of 200 eighth graders, found that 56% of the students had a summer job last year. How many students had a summer job? Show your work.

Wesley
percent • whole = part
$0.56 \cdot 200 = 112$
112 students

Taina
percent • whole = part
$0.56 \cdot 100 = 56$
56 students

Kurano
percent • whole = part
$5.6 \cdot 200 = 1120$
1120 students

GO ON

Guided Practice

Write each percent as a decimal.

1 47%

_____

2 6%

_____

3 60%

_____

4 143%

_____

5 27%

_____

6 9%

_____

7 90%

_____

8 134%

_____

Step by Step Practice

9 Complete the table to show the decimal value, the percent, and the degree measure for each brand of shoe sold at the Fabulous Foot Shoe Store during a recent sale.

Shoe Type	Number Sold	Decimal Value	Percent Value	Degree Measure
Tennis	210			
Skateboard	168			
Basketball	252			
Soccer Cleats	126			
Running/Walking	84			
Total	**840**	**1.00**	**100**	**360°**

Step 1 To write the decimal, divide the number of each type of shoe by the total number of shoes sold.

Step 2 Find the equivalent percent value for each decimal.

Step 3 Change all percent values to degree measures.

Complete the table to show the decimal value, the percent, and the degree measure.

10 Mr. Chiang analyzed the types of advertisements shown on children's TV shows. His team recorded advertisements shown on one Saturday morning.

Ad Type	Number Shown	Decimal Value	Percent Value	Degree Measure
Electronic Games	140			
Breakfast Cereals	40			
Fast Food	100			
Toys	120			
Total	**400**	**1.00**	**100**	**360°**

Step (by) Step *Problem-Solving Practice*

Use the table in Exercise 10 to create a circle graph.

11 **RESEARCH** Mr. Chiang has been asked to present the research findings at a regional meeting. He has decided to make a circle graph.

Problem-Solving Strategies
- ☐ Draw a diagram.
- ☑ Use a table.
- ☐ Work backward.
- ☐ Solve a simpler problem.
- ☐ Look for a pattern.

Understand Read the problem. Write what you know.

Electronic Game advertisements were shown _____ of the time.

Breakfast Cereal advertisements were shown _____ of the time.

Fast Food advertisements were shown _____ of the time.

Toy advertisements were shown _____ of the time.

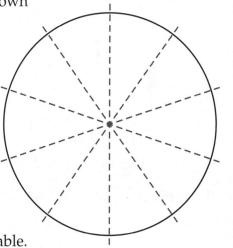

Plan Pick a strategy. One strategy is to use a table.

Solve The title of the graph is _____.

Draw sector lines using 10% marks as guides. Label the sectors of the circle for each category.

Color each sector a different color. Write the percent value in each sector.

Check Add the percentages for each advertisement. Does the sum equal 100%?

GO ON →

12 SALES The store manager at Fabulous Foot shoe store has been asked to present his sales at a store meeting. He has decided to make a circle graph. Use the table in Exercise 9 on page 368 to create a circle graph. Check off each step.

_____ **Understand: I underlined key words.**

_____ **Plan: To solve this problem, I will** _____.

_____ **Solve: The answer is** _____.

_____ **Check: I checked my answer by** _____.

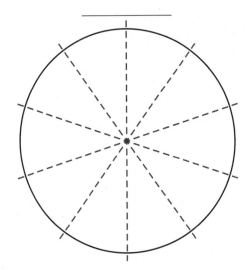

13 Reflect Maisie's table is missing some information. Explain how to use the known percent values to find the number of hours she spent on gymnastics last week.

Activity	Number of Hours	Decimal Value	Percent Value
Homework	15	15 ÷ 30 = 0.50	50%
Piano	3	3 ÷ 30 = 0.10	10%
Gymnastics			
Totals	30	1.00	100

_____

_____

_____

_____

Skills, Concepts, and Problem Solving

Complete the table to show the decimal value, the percent, and the degree measure.

14 Sell-A-Lot car dealership just finished their summer sales drive. The sales team recorded the sales in a table.

Vehicle Type	Number Sold	Decimal Value	Percent Value	Degree Measure
Convertible	104			
SUV	78			
Sedan	52			
Minivan	26			
Total	**260**	**1.00**	**100**	**360°**

15 **SALES** The store manager at Sell-A-Lot wants to present the sales results at the next team meeting. Use the table in Exercise 14 to create a circle graph.

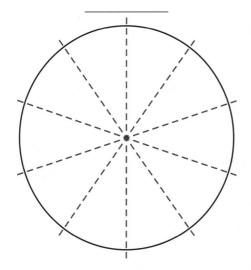

GO ON

16 **FARMER'S MARKET** Complete the table and create a circle graph to show the fruit sold at the local farmer's market.

Type of Fruit	Pounds	Decimal Value	Percent Value	Degree Measure
Apples	81			
Pears	45			
Peaches	36			
Plums	18			
Total	**180**	**1.00**	**100%**	**360°**

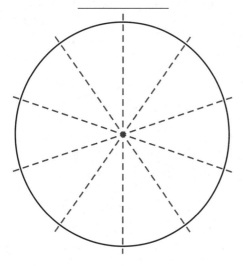

Spiral Review

17 **SURVEYS** The eighth-grade students voted on lunch choices for the last day of school. Find the degrees in each sector. (Lesson 8-2, p. 349)

Lunch Choices

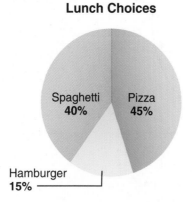

Progress Check 2 (Lessons 8-3 and 8-4)

The Big Buffet restaurant took an inventory of the last 200 meals that were served. Use the circle graph to answer the questions below.

Meals Served

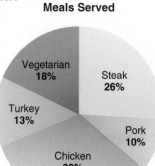

1. What percent of the meals were steak dinners? _____

2. What percent is equal to half of 26%?

   _____ $\cdot \dfrac{1}{2}$ = _____

3. What type of meal was sold half as often as steak dinners?

   _____ were sold half as often as steak dinners.

4. How many meals were chicken or turkey dinners?

   _____

5. How many meals were not vegetarian?

   _____

Complete the table and create a circle graph to show the vegetables sold at the local farmer's market.

6.

Type of Vegetable	Pounds	Decimal Value	Percent Value	Degree Measure
Green Beans	72			
Carrots	90			
Corn	36			
Peas	162			
Total	360	1.00	100%	360°

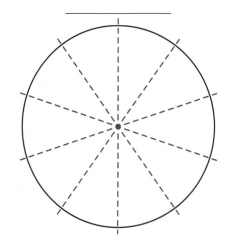

Study Guide

Vocabulary and Concept Check

center, *p. 365*

circle graph, *p. 357*

data, *p. 357*

degree, *p. 349*

denominator, *p. 349*

equivalent fraction, *p. 342*

percent, *p. 342*

ratio, *p. 342*

sector, *p. 349*

simplest form, *p. 349*

Write the vocabulary word that completes each sentence.

1 A ratio that compares a number to 100 is called a(n) _____.

2 _____ are fractions that name the same number.

3 The bottom number in a fraction that represents the number of parts in the whole is called a(n) _____.

4 The most common unit of measure for an angle is a(n) _____.

5 _____ is information which is gathered for statistical purposes.

Label the diagram with the correct vocabulary term.

6 _____

7 _____

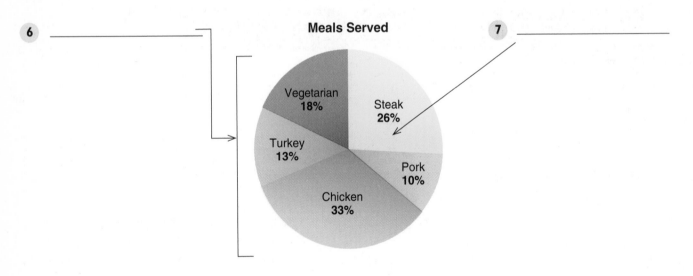

Meals Served

Lesson Review

8-1 Percents (pp. 342–348)

Identify the percent that is modeled.

8

fraction: _____

equivalent fraction: _____

percentage: _____

Example 1

Identify the percent that is modeled.

1. Find the fraction that is shaded.

   $1\frac{7}{10}$ or $\frac{17}{10}$

2. Find an equivalent fraction with a denominator of 100.

   $\frac{17 \cdot 10}{10 \cdot 10} = \frac{170}{100}$

3. Write the percentage.

   170%

8-2 Percents and Angle Measures (pp. 349–355)

Use combinations to find the degrees needed to show each sector in a circle graph.

9

75% = _____ + _____ = 75%

_____ + _____ = _____

10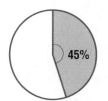

45% = _____ + _____ = 45%

_____ + _____ = _____

Example 2

Use combinations to find the degrees needed to show a 65% sector in a circle graph.

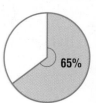

1. Write 65% as a combination of percents.

   50% + 10% + 5% = 65%

2. Show each percent in sector degrees.

   $50\% = \frac{360}{1} \cdot \frac{50}{100} = \frac{18,000}{100} = 180°$

   $10\% = \frac{360}{1} \cdot \frac{10}{100} = \frac{3,600}{100} = 36°$

   $5\% = \frac{360}{1} \cdot \frac{5}{100} = \frac{1,800}{100} = 18°$

3. Find the sum of the sector degrees.

   50% + 10% + 5% = 65%

   180° + 36° + 18° = 234°

8-3 Interpret Circle Graphs (pp. 357–364)

Use the circle graph to answer the questions below.

Last year 300 students took art classes at the museum. At the end of each session students were asked to name their favorite project.

Favorite Project

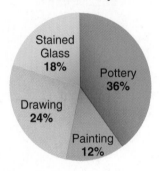

11 What percent of students preferred pottery?

_____

12 What is $\frac{1}{2}$ of 36%?

_____

13 What project was preferred by half as many students as pottery?

_____

14 How many students preferred pottery?

_____

15 How many students preferred painting?

_____

16 How many students did not prefer drawing?

_____

_____

Example 3

Last month 200 visitors from five different states visited the art museum. The number of visitors from which two states was equal to the number of visitors from Florida?

Museum Visitors

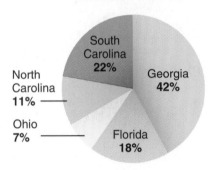

1. What percent of visitors were from Florida? **18%**

2. What two states equal 18% when added together?

   Ohio + North Carolina = 7% + 11% = 18%

   Ohio and North Carolina were represented as often as Florida.

Example 4

How many visitors in the circle graph for Example 3 were from South Carolina?

1. What percent of the visitors were from South Carolina? **22%**

   There were 200 total visitors.

2. Use the percent equation to solve.

   0.22 · 200 visitors = 44 visitors

8-4 Create Circle Graphs (pp. 365–372)

17 Complete the table and create a circle graph to show the Teacher of the Year results.

Name	Number	Percent	Degrees
Mrs. Sato	60		
Ms. Wilson	180		
Mr. Cruz	40		
Mr. Numkena	120		
Total	**400**	**100%**	**360°**

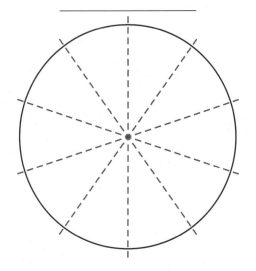

Example 5

Complete the table and create a circle graph to show the type of pets owned.

Pet Adopt, Inc. took a survey of 500 pet owners to see what type of pets they own.

Type of Pet	Number	Percent	Degrees
Dog	300	60%	216°
Horse	25	5%	18°
Cat	100	20%	72°
Fish	50	10%	36°
Bird	25	5%	18°
Total	**500**	**100%**	**360°**

1. To write the decimal, divide the number for each type of pet by the total number of pets.

2. Find the equivalent percent value for each decimal. Change all percent values to degree measures.

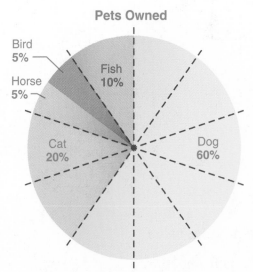

3. Write a title for the graph. Draw sector lines.

4. Label the sectors for each category. Color each sector a different color. Write the percent value for each sector.

Chapter Test

Identify the percent that is modeled.

1

fraction: _____

fraction with a denominator of 100:

$$\frac{\square \cdot \boxed{}}{\square \cdot \boxed{}} = \frac{\square}{100}$$

decimal: _____

percent: _____

2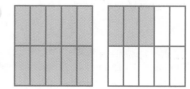

fraction: _____

fraction with a denominator of 100:

$$\frac{\square \cdot \boxed{}}{\square \cdot \boxed{}} = \frac{\square}{100}$$

decimal: _____

percent: _____

Find the degrees needed to show each sector in a circle graph.

3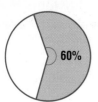

$$60\% = \frac{\square}{100} \div \frac{\boxed{}}{\boxed{}} = \frac{\square}{\square}$$

$$\frac{\square}{\square} \cdot \frac{\square}{\square} = \frac{\square}{\square} = \underline{\quad}$$

4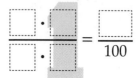

$$10\% = \frac{\square}{100} \div \frac{\boxed{}}{\boxed{}} = \frac{\square}{\square}$$

$$\frac{\square}{\square} \cdot \frac{\square}{\square} = \frac{\square}{\square} = \underline{\quad}$$

Find the missing value.

5 What is 7% of 50?

_____ • _____ = _____

6 What is 45% of 360?

_____ • _____ = _____

7 What is 75% of 350?

_____ • _____ = _____

8 What is 170% of 90?

_____ • _____ = _____

BICYCLES Smithtown Bike Shop has 400 bicycles. Use the circle graph to answer the questions below.

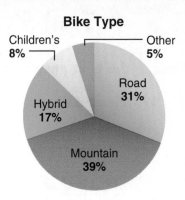

Bike Type

Children's 8%
Other 5%
Road 31%
Hybrid 17%
Mountain 39%

9 How many mountain bikes are at the bike shop?

_____

10 How many road bikes and hybrid bikes are at the bike shop?

_____

11 How many bikes are not children's?

_____

12 Complete the table and create a circle graph to show the boats sold at Boyd's Boat House.

Type of Boat	Number Sold	Decimal Value	Percent Value	Degree Measure
Canoe	104			
Kayak	78			
Motor Boat	52			
Sailboat	26			
Total	260	1.00	100%	360°

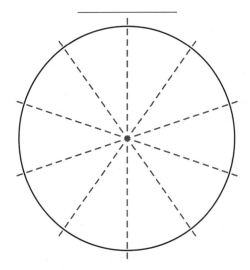

13 **TIPS** Alida paid $17.00 for dinner. She has decided to tip the waiter 15% of the cost of her dinner. How much money will she use to tip the waiter?

_____

STOP

Test Practice

Choose the best answer and fill in the corresponding circle on the sheet at right.

I Which model represents 20%?

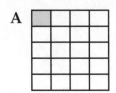

2 What is 28% of 250?

A 28   C 75

B 70   D 700

3 Amanda is making a circle graph of the votes for class president. Forty-five percent of the students voted for Horatio. How many degrees are needed for the sector of votes for Horatio?

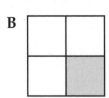

A 45°   C 162°

B 90°   D 360°

4 What is 98% of 380?

A 7.6   C 372.4

B 364.8   D 387.6

5 A large birthday cake was ordered for Patty's birthday. The cake had 75 pieces. Twenty percent of the cake was decorated with roses. How many pieces of cake were decorated with roses?

A 15 pieces   C 75 pieces

B 20 pieces   D 95 pieces

6 In a circle graph, how many degrees are needed to show a 25% sector?

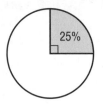

A 25°   C 180°

B 90°   D 360°

7 Smalltown Library checked out 300 books last week. How many of the books were not children's books?

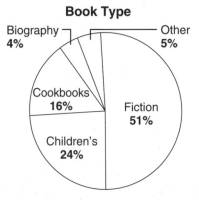

Book Type

Biography 4%
Other 5%
Cookbooks 16%
Fiction 51%
Children's 24%

A 24 books   C 72 books

B 76 books   D 228 books

8 Picture frames are on sale for 25% off the original price. If the cost of a picture frame was $8.00, how much is the discount?

    A $2.00   **C** $6.00

    B $2.50   **D** $10.00

---

9 Which percent represents the shaded portion of the model?

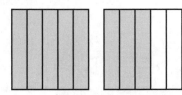

    A 1.35%   **C** 103%

    B 60%   **D** 160%

---

10 Smalltown Library checked out 300 books last week. How many of the books were biographies?

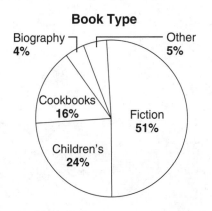

Book Type

Biography 4%
Other 5%
Cookbooks 16%
Fiction 51%
Children's 24%

    A 4 books   **C** 12 books

    B 8 books   **D** 40 books

11 One sector of a circle graph has a measure of 18°. What percent of the circle is represented by this sector?

    A 0.5%   **C** 18%

    B 1.8%   **D** 5%

ANSWER SHEET

Directions: Fill in the circle for each correct answer.

1 (A) (B) (C) (D)
2 (A) (B) (C) (D)
3 (A) (B) (C) (D)
4 (A) (B) (C) (D)
5 (A) (B) (C) (D)
6 (A) (B) (C) (D)
7 (A) (B) (C) (D)
8 (A) (B) (C) (D)
9 (A) (B) (C) (D)
10 (A) (B) (C) (D)
11 (A) (B) (C) (D)

Success Strategy

Read the entire question before looking at the answer choices. Make sure you know what the question is asking.

STOP

Chapter 9

Two-Variable Data

Data is used to compare and predict trends.

Many people compare and predict trends of consumer purchases. Companies often predict purchases based on variables such as weather, season, or popularity of similar products. For example, outdoor markets and roadside vendors often sell more produce on days with pleasant weather conditions than on rainy or stormy days.

STEP 1 Quiz

Math Online > Are you ready for Chapter 9? Take the Online Readiness Quiz at *glencoe.com* to find out.

STEP 2 Preview

Get ready for Chapter 9. Review these skills and compare them with what you will learn in this chapter.

What You Know	What You Will Learn
You know how to solve an equation with one variable.	*Lesson 9-1*

You know how to solve an equation with one variable.

$$16 = 2x$$
$$16 \div 2 = 2x \div 2$$
$$8 = x$$

TRY IT!

Solve for x.

1 $4 + x = 6$ **2** $24 \div x = 8$

_____   _____

Lesson 9-1

You can find **solutions** for equations with two variables.

$$y = 2x$$

$2 = 2 \bullet 1$ If $x = 1$, then $y = 2$.

$4 = 2 \bullet 2$ If $x = 2$, then $y = 4$.

$6 = 2 \bullet 3$ If $x = 3$, then $y = 6$.

The solution for the value of y depends on the value of x.

You know that a coordinate grid has x- and y-coordinates. These coordinates describe the location of a point on the grid.

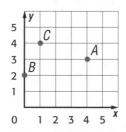

Point A is at (4, 3). To locate Point A, follow the x-axis to the right 4 units. Then follow the y-axis up 3 units.

TRY IT!

Name the location of each point on the coordinate grid.

3 Point B _____

4 Point C _____

Lesson 9-2

A **scatter plot** uses x- and y-axes to show relationships between two sets of **data**.

Scatter plots are graphed as ordered pairs on a coordinate grid.

The Cost of Groceries

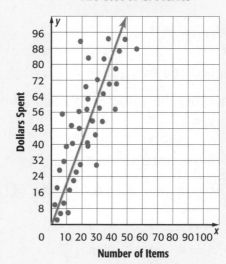

383

Transition to Two-Variable Data

KEY Concept

Solutions for equations with two variables can vary. These solution sets consist of two numbers, one for each variable. Usually, the solution is expressed as an **ordered pair** (x, y).

One-Variable Equation

$$8 = 2x$$

unknown variable

To solve, divide each side by 2.

$$8 = 2x$$
$$\frac{8}{2} = \frac{2x}{2}$$
$$4 = x$$

The solution is 4.

Two-Variable Equation

$$y = 2x$$

unknown variables

To find solutions, make a table.

x	2x	y	(x, y)
0	2·0	0	(0, 0)
1	2·1	2	(1, 2)
2	2·2	4	(2, 4)

Three solutions are $(0, 0)$, $(1, 2)$, and $(2, 4)$.

Finding a solution means finding values for the variables that make the sentence true.

VOCABULARY

ordered pair
a pair of numbers used to locate a point in the coordinate system; the ordered pair is written in this form: (x-coordinate, y-coordinate)

solution
the value of a variable that makes an equation true

x-coordinate
the first number of an ordered pair

y-coordinate
the second number of an ordered pair

Any ordered pair that makes an equation with two variables a true sentence is a solution of the equation. So, an equation with two variables has an infinite number of solutions.

Example 1

Which ordered pair, (3, 6) or (7, 4), is a solution of y = x − 3?

1. Substitute the values for x and y of $(3, 6)$ into the equation.

$$y = x - 3$$
$$6 \stackrel{?}{=} 3 - 3$$
$$6 \neq 0 \; \textit{not a solution}$$

2. Substitute the values for x and y of $(7, 4)$ into the equation.

$$y = x - 3$$
$$4 \stackrel{?}{=} 7 - 3$$
$$4 = 4 \; \checkmark \; \text{a solution}$$

3. The ordered pair $(7, 4)$ is a solution of the equation $y = x - 3$.

YOUR TURN!

Which ordered pair, (−1, −1) or (0, 2), is a solution of $y = 2x + 1$?

1. Substitute the values for x and y of (−1, −1) into the equation.

$$\underline{\hspace{1cm}} \stackrel{?}{=} 2\underline{\hspace{1cm}} + 1$$
$$\underline{\hspace{1cm}} \stackrel{?}{=} \underline{\hspace{1cm}} + 1$$
$$\underline{\hspace{1cm}} \bigcirc \underline{\hspace{1cm}} \qquad \text{Is (−1, −1) a solution?} \underline{\hspace{1cm}}$$

2. Substitute the values for x and y of (0, 2) into the equation.

$$\underline{\hspace{1cm}} \stackrel{?}{=} 2\underline{\hspace{1cm}} + 1$$
$$\underline{\hspace{1cm}} \stackrel{?}{=} \underline{\hspace{1cm}} + 1$$
$$\underline{\hspace{1cm}} \bigcirc \underline{\hspace{1cm}} \qquad \text{Is (0, 2) a solution?} \underline{\hspace{1cm}}$$

3. The ordered pair _____ is a solution of the equation $y = 2x + 1$.

Example 2

**Complete the table for the equation $y = 4x + 5$.
Then find three solutions for the equation.**

1. Make a table. Select three values for x.
   Substitute the values for x in the
   expression $4x + 5$.

2. Complete the table. Find y. Then
   write the ordered pairs.

x	$4x + 5$	y	(x, y)
0	$4(0) + 5$	5	(0, 5)
1	$4(1) + 5$	9	(1, 9)
2	$4(2) + 5$	13	(2, 13)

3. Three solutions of the equation $y = 4x + 5$ are (0, 5), (1, 9), and (2, 13).

YOUR TURN!

**Complete the table for the equation $y = -x + 7$.
Then find three solutions for the equation.**

1. Make a table. Select three values for x.
   Substitute the values for x in the
   expression $-x + 7$.

2. Complete the table. Find y. Then
   write the ordered pairs.

x	$-x + 7$	y	(x, y)

3. Three solutions of the equation $y = -x + 7$ are _____, _____, and _____.

Who is Correct?

Which ordered pair is a solution of $y = x + 12$?

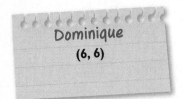
Dominique
(6, 6)

Michael
(1, 11)

Cristina
(3, 15)

Circle the correct answer(s). Cross out incorrect answer(s).

 Guided Practice

Determine which ordered pair, (2, 6) or (3, −1), is a solution of each equation.

1 $y = x - 4$

2 $y = 3x$

3 $y = 2x + 2$

_____

_____

_____

Step by Step Practice

4 Complete the table for the equation $y = 8x$.
Then find three solutions for the equation.

Step 1 Make a table. Select three values for x.
Substitute these values for x in the
expression $y = 8x$.

Step 2 Complete the table. Find y. Then
write the ordered pairs.

x	8x	y	(x, y)

Step 3 Three solutions are _____.

Complete the table for each equation.
Then find three solutions for each equation.

5 $y = -10x$

x	−10x	y
−2	−10(−2)	
6	−10(__)	
7	−10(__)	

6 $y = x + 9$

x	x + 9	y
	__ + 9	
	__ + 9	
	__ + 9	

_____

_____

Find three solutions for each equation.

7 $y = 3x + 2$

8 $y = 5x - 5$

_____

_____

Step by Step Problem-Solving Practice

Solve.

9 **MUSIC** The equation $y = 0.99x$ represents the total cost, y, of downloading x songs from an online music store. Find three solutions of the equation when x equals 2, 5, and 12. Describe what the solutions mean.

Problem-Solving Strategies
- ☐ Draw a diagram.
- ☐ Use logical reasoning.
- ☐ Solve a simpler problem.
- ☐ Work backward.
- ☑ Make a table.

Understand   The variable x represents _____.

The variable y represents _____.

The values of x are _____.

Plan   Pick a strategy. One strategy is to make a table.

Solve

x	$0.99x$	y	(x, y)

The ordered pair solutions are _____.

This means that the cost of downloading 2 songs is _____.

The cost of downloading 5 songs is _____.

The cost of downloading 12 songs is _____.

Check   Use estimation to check your answer. One song costs about \$1, two songs cost about \$2, and so on.

GO ON

10 **PETS** The equation $y = 14 - x$ represents the number of puppies in a pet store, y, after x puppies are sold. Find three solutions of the equation when x equals 6, 10, and 13.

Check off each step.

_____ Understand: I underlined key words.

_____ Plan: To solve the problem, I will _____.

_____ Solve: The answer is _____.

_____ Check: I checked my answer by _____.

_____.

Solve.

11 **MONEY** The equation $y = 6x + 5$ represents the amount of money Marissa has, y, after baby-sitting x hours. Find three solutions of the equation. Describe what the solutions mean.

_____

_____

_____

12 **CYCLING** The equation $d = 7.5t$ represents the distance, d, a cyclist travels after t hours. Find three solutions of the equation. Describe what the solutions mean.

_____

_____

_____

_____

13 **Reflect** The sum of Nate's and Ling's ages is 25. Let x represent Nate's age and let y represent Ling's age. Find four ordered pairs of numbers that represent their possible ages.

_____

▶ Skills, Concepts, and Problem Solving

Determine which ordered pair, (0, 0), (5, −9), or (1, 3), is a solution of each equation.

14 $y = 7x$

_____

15 $y = x + 2$

_____

16 $y = 5x - 2$

_____

17 $y = -x - 4$

_____

Complete the table for each equation.
Then find three solutions for each equation.

18 $y = 2x$

x	2x	y
0	2( )	
5	2( )	
9	2( )	

_____

19 $y = x + 15$

x	x + 15	y
−1	−1 +	
3	__	
6	__	

_____

20 $y = 6x - 5$

x	6x − 5	y
	__	
	__	
	__	

_____

21 $y = -x + 3$

x	−x + 3	y
	__	
	__	
	__	

_____

Find three solutions for each equation.

22 $y = x$

_____

23 $y = -4x$

_____

GO ON

Find three solutions for each equation.

24 $y = x + 11$

_____

25 $y = x - 5$

_____

26 $y = 4x + 2$

_____

27 $y = 10x - 3$

_____

Solve.

28 **WATER** It costs a city about $10 per minute when a fire hydrant is running. This can be represented by the equation $C = 10m$, where C is the cost and m is the number of minutes. Find three solutions of the equation. Choose one solution and describe what it means.

_____

_____

29 **GEOMETRY** The formula for the area of a rectangle is $A = \ell \cdot w$. The area of a rectangle is 48 square centimeters. Find three solutions of the equation. Choose one solution and describe what it means.

_____

_____

Vocabulary Check **Write the vocabulary word that completes each sentence.**

30 A pair of numbers used to locate a point in the coordinate system is called a(n) _____.

31 The ordered pair (1, 7) is a(n) _____ of the equation $y = 2x + 5$.

32 **Writing in Math** What is the solution of the equation $3x + y = 14$ if $y = 2$? Explain your reasoning.

_____

_____

STOP

Scatter Plots

KEY Concept

In a **scatter plot**, two sets of data are graphed as ordered pairs on a **coordinate grid**. A scatter plot shows the relationship, if any relationship exists, between the two sets of data.

Positive Relationship

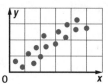

As *x* increases, *y* increases.

Negative Relationship

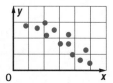

As *x* increases, *y* decreases.

No Relationship

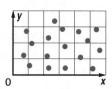

There is no obvious pattern.

If the points go up from left to right, the data show a positive relationship.

If the points go down from left to right, the data show a negative relationship.

If the points show no pattern from left to right, the data show no relationship.

Copyright © Glencoe/McGraw-Hill, a division of The McGraw-Hill Companies, Inc.

Like line graphs, scatter plots are useful for interpreting data and making predictions. They often show trends in data.

GO ON

Example 1

Explain whether the scatter plot shows a *positive*, *negative*, or *no* relationship.

1. Describe the two sets of data shown in the scatter plot.

   The two sets of data are the ages and the weights of thirty horses that are less than a year old.

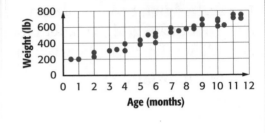

Growth of Horses

2. What do the *x*-axis and the *y*-axis represent?

   The *x*-axis represents the ages of the horses.
   The *y*-axis represents the weights of the horses.

3. In general, as the ages of the horses increase, what happens to the weights of the horses?

   The weights increase.

4. The scatter plot shows a positive relationship between the ages and weights of the horses.

YOUR TURN!

Explain whether the scatter plot shows a *positive*, *negative*, or *no* relationship.

1. Describe the two sets of data shown in the scatter plot.

   The two sets of data are the _____ twenty students and the _____

   _____.

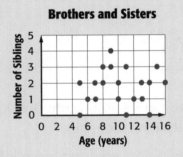

Brothers and Sisters

2. What do the *x*-axis and the *y*-axis represent?

   The *x*-axis represents the _____.
   The *y*-axis represents the _____.

3. As the ages of the students increase, what happens to the number of siblings they have?

   _____

4. The scatter plot shows _____ relationship between the ages of the students and the number of siblings they have.

Example 2

Use the information in the table to create a scatter plot.

Create a scatter plot to show how the amount of snowfall affects the sales of sleds.

1. Set Snowfall (in.) on the *x*-axis.

2. Set Number of Sleds Sold on the *y*-axis.

3. Plot the points from the table.

Selling Sleds

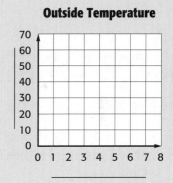

Snowfall (in.)	0	0	1	2	3	3	4	4	5	6	6	7	7	8	9
Number of Sleds Sold	0	2	0	1	3	5	8	10	12	13	16	12	17	15	18

YOUR TURN!

Use the information in the table to create a scatter plot.

Create a scatter plot to show how the changes in outside temperature change over the course of seven hours.

1. Set _____ on the *x*-axis.

2. Set _____ on the *y*-axis.

3. Plot the points from the table.

Outside Temperature

Number of Hours	1	2	3	4	5	6	7
Temperature (°F)	50	48	46	45	43	40	38

Who is Correct?

Explain whether the scatter plot "Selling Sleds" shows a *positive*, *negative*, or *no* relationship.

Addison
The points go down from left to right: negative relationship.

Tamyra
The points go up from left to right: positive relationship.

Carlos
The points show no pattern: no relationship.

Circle the correct answer(s). Cross out incorrect answer(s).

GO ON

Guided Practice

Explain whether each scatter plot shows a *positive*, *negative*, or *no* relationship.

1 Height of Students and Number of CDs

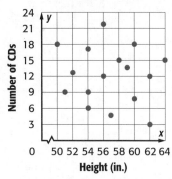

As the height increases, the number of CDs _____.

Describe the relationship between the height of students and the number of CDs that they own.

_____

2 Lemonade Sales

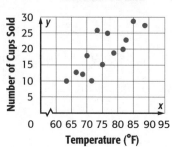

As the temperature increases, the number of cups sold _____.

Describe the relationship between the temperature and the number of cups sold.

_____

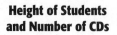 **Step** (by) **Step Practice**

Use the information in the table to create a scatter plot.

3 Create a scatter plot to show how the number of students in the Extreme Frisbee Club has changed over the course of six years.

Step 1 Set _____ on the *x*-axis.

Step 2 Set _____ on the *y*-axis.

Step 3 Plot the points from the table.

Extreme Frisbee

Year	'03	'04	'05	'06	'07	'08	'09
Number of Students	4	9	13	16	23	25	30

Use the information in the table to plan and create a scatter plot.

Year	'03	'04	'05	'06	'07	'08	'09
Number of Students	60	50	39	32	28	25	20

4 Set _____ on the *x*-axis.

5 Set _____ on the *y*-axis.

6
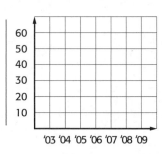
_____

Step (by) Step *Problem-Solving Practice*

Solve.

7 **ROLLER COASTERS** The table below shows the data. Create a scatter plot to find the relationship between the data sets.

Problem-Solving Strategies
☐ Draw a diagram.
☐ Use logical reasoning.
☐ Solve a simpler problem.
☐ Work backward.
☑ Make a graph.

Height (ft.)	120	125	125	140	145	155	165	180	205	215
Top Speed (mph)	52	48	60	60	60	70	67	73	74	78

Understand Read the problem. Write what you know.

The *x*-axis represents the

_____.

The *y*-axis represents the

_____.

Roller Coasters

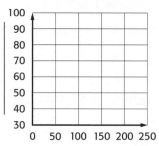

Plan Pick a strategy. One strategy is to make a graph.

Solve The relationship between the height and top speed of the roller coasters is _____.

_____

Check Since the points go up from left to right, the data show a _____ relationship.

GO ON

8 PENCILS Create a scatter plot to show the number of weeks that 20 pencils are used and the pencil lengths. Is the relationship *positive, negative,* or is there *no* relationship? Check off each step.

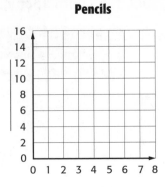
Pencils

Time Used (Weeks)	0	0	1	1	2	2	3	3	3	4
Length (cm)	14	16	14	15	14	13	13	12	10	11

Time Used (Weeks)	4	$4\frac{1}{2}$	5	5	$5\frac{1}{2}$	6	6	6	$6\frac{1}{2}$	7
Length (cm)	10	8	9	7	7	8	7	6	6	7

_____ **Understand:** I underlined key words.

_____ **Plan:** To solve the problem, I will _____.

_____ **Solve:** The answer is _____.

_____ **Check:** I checked my answer by _____.

9 Reflect A scatter plot shows the number of people in a family and their weekly grocery bill. Would you expect the graph of the data to show a positive, negative, or no relationship? Explain.

_____

_____

 Skills, Concepts, and Problem Solving

Explain whether each scatter plot shows a *positive, negative,* or *no* relationship.

10 As the ages increase, the weights _____.

Describe the relationship between the ages and weights of the gazelles.

_____

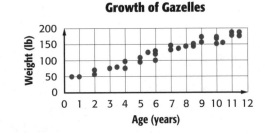
Growth of Gazelles

11 As the temperature increases, the number of fans sold _____.

Describe the relationship between the temperature and fan sales.

_____

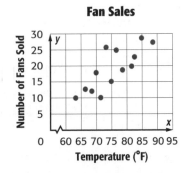
Fan Sales

12 **DROUGHT** A drought affects the depth of the pond. The table below shows the data. Create a scatter plot to find the relationship between the data sets.

Depth of Pond

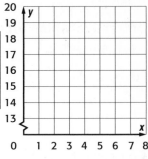

Day	1	2	3	4	5	6	7	8
Depth (in.)	18.5	18	17.5	17.5	17.0	17.0	16.0	16.0

Describe the relationship between the number of days and depth of the pond.

_____

Vocabulary Check **Write the vocabulary word that completes each sentence.**

13 A _____ displays two sets of data on the same graph.

14 If the points on a scatter plot appear to go downhill from left to right, then the data points show a _____ relationship.

15 **Writing in Math** Describe how you can use a scatter plot to display two sets of related data.

_____

_____

 Spiral Review (Lesson 9-1, p. 384)

Determine which ordered pair, (0, −5), (2, 10), or (−3, 15), is a solution of each equation.

16 $y = x + 8$

17 $y = -5x$

18 $y = 3x - 5$

_____   _____   _____

Solve.

19 **FOOD** The equation $C = 2 + 1.5t$ represents the cost of a meal that includes a beverage and t tacos. Find three solutions of the equation. Describe what the solutions mean.

_____

_____

 STOP

Determine which ordered pair, (0, –4), (1, 5), or (4, 3), is a solution of each equation.

1 $y = x + 4$

2 $y = x - 1$

3 $y = 2x - 4$

_____

_____

_____

Complete the table for each equation. Then find three solutions for each equation.

4 $y = 5x$

x	5x	y
1	5(1)	
4		
8		

_____

5 $y = 3x + 1$

x	3x + 1	y
−1		
0		
6		

_____

Find three solutions of each equation.

6 $y = 12x$

7 $y = x - 8$

8 $y = 7x + 2$

_____

_____

_____

Solve.

9 **MEASUREMENT** The equation $P = 6s$ represents the perimeter of a regular hexagon with side length s. Find three solutions of the equation. Describe what the solutions mean.

_____

_____

_____

Determine whether the scatter plot shows a *positive*, *negative*, or *no* relationship.

10

Checking Out Library Items

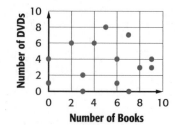

Lines of Best Fit

KEY Concept

If the points on a **scatter plot** come close to lying on a straight line, a **line of best fit** can be drawn to show the trend in the **data**. The line should be very close to most of the data points.

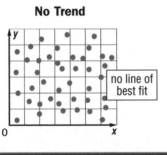

Upward Trend

positive slope

Downward Trend

negative slope

No Trend

no line of best fit

VOCABULARY

data
    information, often numerical, which is gathered for statistical purposes

line of best fit
    a line that is very close to most of the data points on a scatter plot

scatter plot
    in a scatter plot, two sets of related data are plotted as ordered pairs on the same graph

slope
    the rate of change between any two points on a line; the ratio of vertical change to horizontal change

A line of best fit is useful for showing trends in data and for making predictions. Since a line of best fit is an estimate, it is possible to draw slightly different lines for the same data.

Example 1

The scatter plot shows the length and weight of thirteen whale sharks. Draw a line of best fit. Then describe the slope of the line and the trend in the data.

1. Draw a line so that it runs through the center of the group of points.

2. The slope of the line of best fit is positive.

    It shows that there is an upward trend in the weight of whale sharks as the length increases.

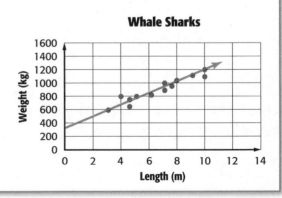

Whale Sharks

GO ON

YOUR TURN!

The scatter plot shows the number of pizzas that the school cafeteria sold during one lunch period. Draw a line of best fit. Then describe the slope of the line and the trend in the data.

1. Draw a line so that it runs through the center of the group of points.

2. The slope of a line of best fit is _____.

   It shows that there is a(n) _____ trend in the number of pizzas sold as the number of minutes increases.

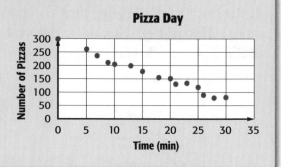

Pizza Day

Example 2

Use the line of best fit to predict the weight of a whale shark that is 13 meters long.

1. Use a ruler to extend the line.

2. Locate the point on the line that has an *x*-value of 13. Find the corresponding *y*-value.

   The corresponding *y*-value is about 1,500.

3. So, a whale shark that is 13 meters long would weigh about 1,500 kilograms.

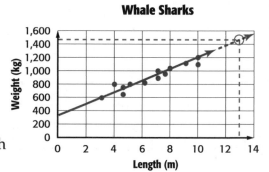

Whale Sharks

YOUR TURN!

Use the line of best fit shown to predict the length of the wait time for the ride at 9:30 P.M.

1. Use a ruler to extend the line.

2. Locate the point on the line that has an *x*-value of _____. Find the corresponding *y*-value.

   The corresponding *y*-value is about _____.

3. So, at 9:30, the wait time will be about

   _____.

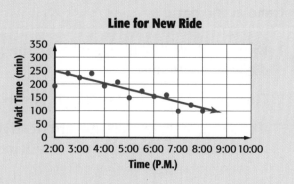

Line for New Ride

Who is Correct?

Describe the slope of the line and the trend in the data in the graph, "Line for New Ride."

Desmond
positive slope
upward trend

Martina
negative slope
no trend

Cassie
negative slope
downward trend

Circle the correct answer(s). Cross out incorrect answer(s).

▶ Guided Practice

Draw a line of best fit. Then describe the slope of the line and the trend in the data.

1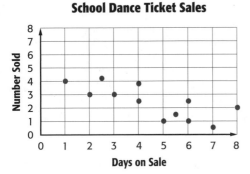
School Dance Ticket Sales

Draw a line of best fit. The line has a _____ slope.

The line shows a(n) _____ trend.

2
Selling Magazine Subscriptions

Draw a line of best fit. The line has a _____ slope.

The line shows a(n) _____ trend.

3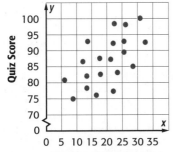
Study Time and Quiz Scores

Draw a line of best fit. The line has a _____ slope.

The line shows a(n) _____ trend.

4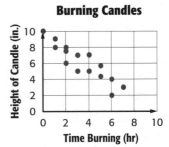
Burning Candles

Draw a line of best fit. The line has a _____ slope.

The line shows a(n) _____ trend.

GO ON

Use the line of best fit to predict how many pairs of new shoes a 15 year old will have.

5 The scatter plot shows the ages of fifteen children and the number of pairs of new shoes that they own.

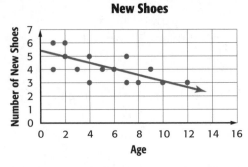

New Shoes

Step 1 Use a ruler to extend the line so that you can predict the *y*-value for an *x*-value of _____.

Step 2 Locate the point on the line that has an *x*-value of _____. Then move across to the *y*-axis to find the corresponding *y*-value.

The corresponding *y*-value is _____.

Step 3 According to the scatter plot, a 15 year old will have about _____ pairs of new shoes.

The scatter plot shows the number of blocks that students walked while participating in a walkathon.

6 Draw a line of best fit. Then describe the slope of the line and the trend in the data.

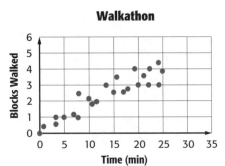

Walkathon

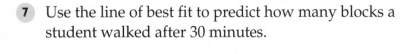

7 Use the line of best fit to predict how many blocks a student walked after 30 minutes.

_____

Solve.

8 **SOCCER** The scatter plot shows the number of goals and the number of assists made by soccer players in a season. Use the line of best fit shown to predict how many assists a soccer player will have if she scores 25 goals.

Understand Read the problem. Write what you know.

The points on the scatter plot represent the _____ and _____.

The line of best fit shows a(n) _____ in the number of assists as the number of goals increases.

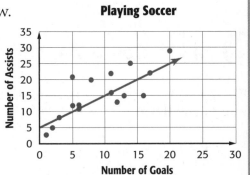

Playing Soccer

Plan Pick a strategy. One strategy is to use a graph. Use a ruler to extend the line.

Solve When the *x*-value is _____, the corresponding *y*-value is about _____.

So, if a soccer player scores 25 goals, you can predict that she will have about _____ assists.

Check Check your answer by plotting the point that you named on the scatter plot.

---

9 **WHITE WATER RAFTING** The scatter plot shows the distances traveled by ten rafts on a river. Use the line of best fit to predict the total distance a raft will travel if it is on the river for 6.5 hours. Check off each step.

_____ Understand: I underlined key words.

_____ Plan: To solve the problem, I will _____.

_____ Solve: The answer is _____.

_____ Check: I checked my answer by _____

_____.

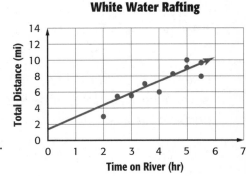

White Water Rafting

GO ON

10 **Reflect** The scatter plot shows the amount of rain that fell during the first ten days of the month. Explain whether you can draw a line of best fit for the data.

_____

_____

_____

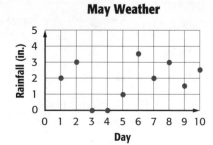

May Weather

Skills, Concepts, and Problem Solving

Draw a line of best fit. Then describe the slope of the line and the trend in the data.

11

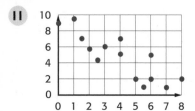

_____

12

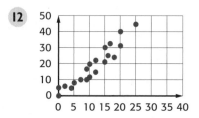

_____

Use the line of best fit to make predictions.

13

Lizards in Pet Store

Use the line of best fit to predict how many lizards will be in the pet store on the 12th day.

_____

14

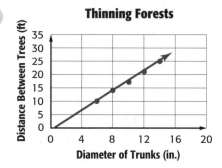

Thinning Forests

Use the line of best fit to predict the distance between trees if the trunks have a diameter of 16 inches.

_____

Solve.

15 PLANTS The scatter plot shows the height of plants that were planted from seeds. Draw a line of best fit. Then describe the slope of the line and the trend in the data.

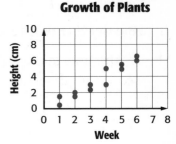

Growth of Plants

_____

16 Use the line of best fit that was drawn to predict the height of a plant after 8 weeks.

_____

Vocabulary Check **Write the vocabulary word that completes each sentence.**

17 A _____ is a line that is very close to most of the data points on a scatter plot.

18 If the slope of a line of best fit is positive, then the points on the scatter plot show a(n) _____ trend.

19 Writing in Math Explain how to use a line of best fit to make a prediction.

_____

_____

_____

▶ Spiral Review

The scatter plot shows the costs of used cars. (Lesson 9-2, p. 391)

20

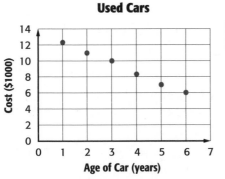

As the age of the car increases, the cost of the car _____.

Describe the relationship between the age of the car and its cost.

_____

STOP

Use the scatter plot to the right to complete Exercises 1–4.

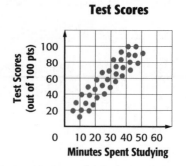

Test Scores

1. As the time spent studying increases, do the test scores increase, decrease, or show no pattern?

   _____

2. Is there a relationship between the time spent studying and test scores? If so, describe the relationship.

   _____

3. Draw a line of best fit on the scatter plot.

4. Use the line of best fit to predict the test score of a student who studies for 30 minutes.

   _____

Use the scatter plot to the right to complete Exercises 5–8.

5. As the number of meals eaten at home increases, does the number of meals eaten at a restaurant increase, decrease, or show no pattern?

   _____

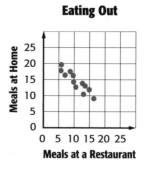

Eating Out

6. Is there a relationship between the number of meals eaten at home, and number of meals eaten at a restaurant? If so, describe the relationship.

   _____

7. Draw a line of best fit on the scatter plot.

8. Use the line of best fit to find the number of meals eaten at home if there are 20 meals eaten at a restaurant.

   _____

Vocabulary and Concept Check

coordinate grid, *p. 391*

data, *p. 399*

line of best fit, *p. 399*

ordered pair, *p. 384*

scatter plot, *p. 391*

slope, *p. 399*

solution, *p. 384*

x-axis, *p. 391*

x-coordinate, *p. 384*

y-axis, *p. 391*

y-coordinate, *p. 384*

Write the vocabulary word that completes each sentence.

1 A(n) _____ is the first number of an ordered pair.

2 Information which is gathered for statistical purposes is called _____.

3 A(n) _____ is the value of a variable that makes an equation true.

4 The vertical line of the two perpendicular number lines in a coordinate grid is called the _____.

5 A(n) _____ is a grid in which a horizontal number line and a vertical number line intersect at their zero points.

Lesson Review

9-1 Transition to Two-Variable Data (pp. 384–390)

6 Which ordered pair, (3, 9) or (1, 6), is a solution of $y = x + 5$?

_____

7 Which ordered pair, (2, 8) or (4, 10), is a solution of $y = 3x + 2$?

_____

8 Which ordered pair, (2, −3) or (0, −5), is a solution of $y = 5x − 5$?

_____

Example 1

Which ordered pair, (5, 4) or (1, 3), is a solution of $y = x − 1$?

1. Substitute the *x*- and *y*-values of (5, 4) into the equation.

$y = x − 1$
$4 \stackrel{?}{=} 5 − 1$
$4 = 4$ ✓ a solution

2. Substitute the *x*- and *y*-values of (1, 3) into the equation.

$y = x − 1$
$3 \stackrel{?}{=} 1 − 1$
$3 \neq 0$ not a solution

Find three solutions for each equation.

9 $y = x + 4$

_____

10 $y = 3x - 2$

_____

11 $y = 2x + 7$

_____

12 $y = -x - 2$

_____

13 $y = -2x + 3$

_____

9-2 Scatter Plots (pp. 391–397)

Explain whether each scatter plot shows a *positive*, *negative*, or *no* relationship.

14

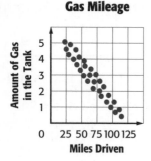

Gas Mileage

_____

15

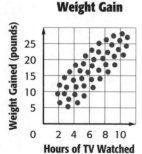

Weight Gain

_____

Example 2

Find three solutions for the equation $y = 2x + 1$.

1. Make a table. Select three values for x. Substitute the values for x in the expression $2x + 1$.

2. Complete the table. Find y. Then write the ordered pairs.

x	$2x + 1$	y	(x, y)
0	$2(0) + 1$	1	$(0, 1)$
1	$2(1) + 1$	3	$(1, 3)$
2	$2(2) + 1$	5	$(2, 5)$

3. Three solutions of the equation $y = 2x + 1$ are $(0, 1)$, $(1, 3)$, and $(2, 5)$.

Example 3

Explain whether the scatter plot below shows a *positive*, *negative*, or *no* relationship.

The scatter plot shows the number of hours worked and the number of calls received by a customer service center.

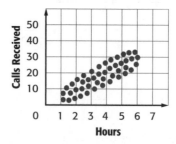

Customer Service Calls

1. The x-axis represents the number of hours worked. The y-axis represents the calls received.

2. In general, as the number of hours worked increases, the calls received increases.

3. The scatter plot shows a positive relationship between the hours worked and the calls received.

Copyright © Glencoe/McGraw-Hill, a division of The McGraw-Hill Companies, Inc.

9-3 Lines of Best Fit (pp. 399–405)

Draw a line of best fit for each scatter plot. Then describe the slope of the line and the trend in the data.

16
Cookies Baked

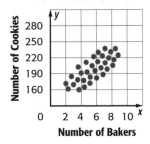

17
Run Times

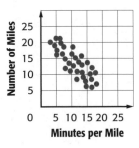

18 Use the line of best fit drawn in Exercise 16 to predict how many cookies 10 bakers will bake.

19 Use the line of best fit drawn in Exercise 17 to predict how many miles can be completed if it takes someone 20 minutes to run a mile.

Example 5

Draw a line of best fit for each scatter plot. Then describe the slope of the line and the trend in the data.

The scatter plot shows the height and weight of the young men in a middle school.

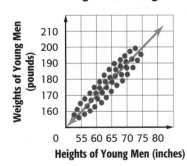

Heights and Weights

1. Draw a line so that it runs through the center of the group of points.

2. The slope of the line of best fit is positive. It shows that there is an upward trend in the young mens' weights as their heights increase.

Example 6

Use the line of best fit drawn in Example 5. Predict how much a young man who is 76 inches tall would weigh.

1. Use a ruler to extend the line.

2. Locate the point on the line with an x-value of 76. Then move across to the y-axis to find the corresponding y-value.

   The corresponding y-value is about 205.

3. So, a young man that is 76 inches tall would weigh about 205 pounds.

Complete the table for each equation. Then find three solutions for each equation.

1 $y = x - 2$

x	x − 2	y
1		
3		
5		

_____

2 $y = -3x + 2$

x	−3x + 2	y
−2	____	
2		
4	____	

_____

Determine which ordered pair, (1, 2) or (2, 8), is a solution of each equation.

3 $y = x + 6$

_____

4 $y = 2x$

_____

5 $y = 3x + 2$

_____

6 $y = -2x + 4$

_____

Explain whether each scatter plot shows a _positive_, _negative_, or _no_ relationship.

7 **Chili Sales**

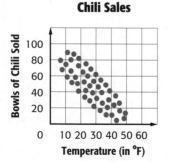

As the temperature increases, the number of bowls of chili sold _____.

Describe the relationship between the temperature and the number of bowls of chili sold.

_____

8 **Backyard Gardens**

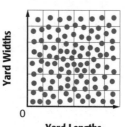

As yard lengths increase, the yards widths _____.

Describe the relationship between the yard length and yard width.

_____

Use the scatter plot of the number of students who drove to school from 2004 to 2009 to complete Exercises 9–10.

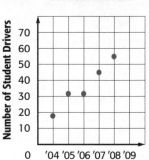

Student Drivers

9 Do the data in the scatter plot show a *positive*, *negative*, or *no* relationship? Explain.

_____

_____

10 Draw a line of best fit. Predict the number of students who will drive to school in 2009.

_____

Solve.

11 **CELL PHONES** A local cell phone company charges a connection fee of $10. The company also charges 2 cents per minute for all calls. The equation $y = 0.02x + 10$ represents the total monthly cost (y) of using x minutes of air time. Find three solutions of the equation when $x = 10$, 100, and $1,000$. Explain the solutions.

_____

_____

12 The scatter plot shows the number of points scored and the number of assists made by basketball players in one season. Draw a line of best fit. Predict how many assists a basketball player will have if he scores 40 points.

_____

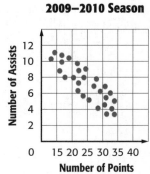

2009–2010 Season

Correct the mistakes.

13 The equation $y = -2x + 1$ was written on the board. The teacher asked the class to give the solution if $x = -1$. Ken answered $(-1, 3)$, and Mila answered $(-1, -1)$. Who is correct? What mistake was made?

_____

_____

Choose the best answer and fill in the corresponding circle on the sheet at the right.

1 Which of the following best describes the relationship between the number of tickets and miles per hour over the speed limit in this scatter plot?

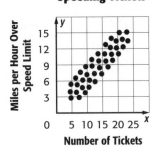

Speeding Tickets

A positive relationship

B negative relationship

C no relationship

D unknown

2 Which is a solution of the equation $y = -2x + 5$?

A $(0, 3)$

B $(7, 0)$

C $(-1, 7)$

D $(1, -7)$

3 Find two solutions of the equation $y = x - 2$.

A $(1, -3), (2, -4)$

B $(3, 5), (0, 2)$

C $(-2, 0), (3, 5)$

D $(0, -2), (5, 3)$

4 Which of the following scatter plots shows no relationship between x and y?

A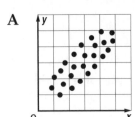

B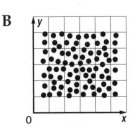

C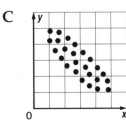

D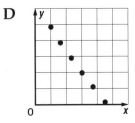

5 Find three solutions of the equation $y = 2x + 3$.

A $(0, 3), (1, 5), (2, 7)$

B $(0, 3), (1, 1), (2, 7)$

C $(2, 3), (3, 4), (4, 5)$

D $(0, 3), (2, 3), (4, 11)$

6 (5, 10) is a solution to which of the following equations?

   A $y = -2x$

   B $y = x + 2$

   C $y = x - 2$

   D $y = 2x$

---

7 Which is a solution to the equation $y = -5x - 5$?

   A $(-2, -15)$

   B $(-2, 15)$

   C $(2, -15)$

   D $(2, 5)$

---

8 Which of the following best describes the relationship between age and amount of water consumed per day in this scatter plot?

Drinking Water Habits

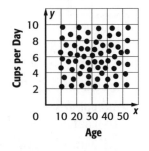

   A positive relationship

   B negative relationship

   C no relationship

   D unknown

9 The scatter plot shows the temperature at various times of one day. Use a line of best fit to predict the temperature at 2:00 P.M.

Hourly Temperatures

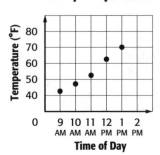

   A 40°   **C** 80°

   B 70°   **D** 90°

ANSWER SHEET

Directions: Fill in the circle of each correct answer.

1   (A) (B) (C) (D)

2   (A) (B) (C) (D)

3   (A) (B) (C) (D)

4   (A) (B) (C) (D)

5   (A) (B) (C) (D)

6   (A) (B) (C) (D)

7   (A) (B) (C) (D)

8   (A) (B) (C) (D)

9   (A) (B) (C) (D)

Success Strategy

After answering all the questions, go back and check the signs and operations, making sure you worked each problem correctly.

Index